"Ann Dowsett Johnston brilliantly captures how addiction stealthily invades lives. She courageously delves into the lure of alcohol—whether it comes from the culture around us, the false promises, or the pain we seek to numb. *Drink* is an important and timely call to attention for the rise of binge-drinking and alcohol addiction in women of all ages."

—David A. Kessler, MD, author of the *New York Times* bestseller *The End of Overeating* and former commissioner of the FDA

"Johnston brings the weight of her journalism and academic experience to build a convincing case that women are increasingly succumbing to the dark side of alcohol. The book is meant to alarm us, one searing fact at a time."

—*Washington Post*

"Veteran journalist Dowsett Johnston combines her powerful personal story with strong reporting to show that alcohol harms women even more than it does men. . . . A powerful case for drinking water, not wine."

—*Booklist*

"A compelling sociological study and memoir."   —*Kirkus Reviews*

"*Drink* is a skilled combination of memoir and reporting."   —*Globe and Mail*

"In this hauntingly written book, Dowsett Johnston writes that alcohol has become 'the modern woman's steroid.' Ouch. Feminists may push back (especially those who like a little Pinot with their equality). But I think she gets it right. Women are working harder and drinking harder. . . . *Drink* is a wonderful book, written by a woman who could be your best friend."

—Judith Timson, *Toronto Star*

"This brave book, half memoir and half meticulous research, is one woman's story of her spiraling alcohol dependency and her fearless look at a growing and destructive trend afflicting women of all demographics."

—*Cosmopolitan*

"A powerful and important book about the increase in alcoholism and binge-drinking among women, and about our willful blindness to the damages of drinking in our culture."

—Susan Cheever, author of *My Name Is Bill: Bill Wilson— His Life, and the Creation of Alcoholics Anonymous*

"In this comprehensively researched and insightful book, Ann Dowsett Johnston chronicles her own destructive dance with alcohol and her recovery, and explores disturbing trends in contemporary women's relationship with alcohol. A crucially important book for anyone interested in women's health and addiction issues."

—Susan Juby, author of *Nice Recovery*

"*Drink* is a gift to women, to parents, and to all who want to understand the experience of alcoholism. The writing is gripping and vivid, the voice personal, the research exacting, the stories revealing if sometimes heartbreaking, the conclusions essential. A triumphant life, a triumphant book."

—Gabor Maté, MD, author of *In the Realm of Hungry Ghosts: Close Encounters with Addiction*

"Part elegiac memoir and part synthetic journalism, this book bluntly tackles the silence and complicity about alcoholism in women. It traverses subjects from brain science to public policy, from trauma to culture, always with powerful stories of women for whom alcohol became an insidious and destructive force. Without soft-pedaling the damage and losses, it is also a story of love, triumph, and hope."

—David Goldbloom, MD, FRCPC, senior medical adviser, Centre for Addiction and Mental Health, Toronto

"A game-changing look at one of our culture's hidden problems. . . . Honest, brave, and inspirational."

—Margaret Trudeau, author of *Changing My Mind*

Drink

The Intimate Relationship Between
Women and Alcohol

ANN DOWSETT JOHNSTON

HARPER WAVE
An Imprint of HarperCollinsPublishers
www.harperwave.com

HarperCollins books may be purchased for educational, business, or sales promotional use. For information, please e-mail the Special Markets Department at SPsales@harpercollins.com.

A hardcover edition of this book was published in 2013 by HarperWave, an imprint of HarperCollins Publishers.

FIRST HARPER WAVE PAPERBACK EDITION PUBLISHED 2014

Designed by Leah Carlson-Stanisic

Library of Congress Cataloging-in-Publication Data has been applied for.

ISBN 978-0-06-224180-1

17 18   OV/RRD   10 9 8 7 6 5

AUTHOR'S NOTE

The names and other identifying details of some major and minor characters have been changed to protect individual privacy and anonymity.

TO MY MOTHER,
for her courage and love
AND TO NICHOLAS,
for his infinite wisdom

*Our excesses are the best clue we have to our own poverty,
and our best way of concealing it from ourselves.*

—ADAM PHILLIPS, BRITISH PSYCHOANALYST

the laughing heart

by Charles Bukowski

your life is your life
don't let it be clubbed into dank
submission.
be on the watch.
there are ways out.
there is a light somewhere.
it may not be much light but
it beats the
darkness.
be on the watch.
the gods will offer you
chances.
know them, take them.
you can't beat death but
you can beat death
in life,
sometimes.
and the more often you
learn to do it,
the more light there will
be.
your life is your life.
know it while you have
it.
you are marvelous.
the gods wait to delight
in
you.

CONTENTS

Prologue

Hang out in the brightly lit rooms of AA, or in coffee shops, talking to dozens of women who have given up drinking, and this is the conclusion you come to: for most, booze is a loan shark, someone they trusted for a while, came to count on, before it turned ugly.

Every person with a drinking problem learns this the hard way. And no matter what the circumstances, certain parts of the story are always the same. Here is how the story goes:

At first, alcohol is that elegant figure standing in the corner by the bar, the handsome one in the beautiful black tuxedo. Or maybe he's in black leather and jeans. It doesn't matter. You can't miss him. He's always at the party—and he always gets there first.

Maybe you first saw him in high school. Many do. Others meet him long before. He finds his moment, some time when you're wobbly or nervous, excited or scared. You're heading into a big party or a dance. All of a sudden your stomach begins to lurch. You're overdressed, or underdressed; too tall, too short; heartsick, or heart-in-your-mouth anxious. Doesn't matter. Booze wastes no time. He sidles up with a quick hit of courage. You grab it. It feels good. It works.

Or maybe you've fallen in love. You're at a wedding, a dinner, a celebration. You want this moment to last. You fear it won't. Just as your doubts begin to get the best of you, booze holds out a glass, a slim stem of liquid swagger, pale blond and bubbly. You take a sip and instantly the room begins to soften. So do you: your toes curl a little, your heart is light. All things are possible. Now *this* is a sweetheart deal.

This is how it begins. And for many, this is where it ends. Turning twenty-one or twenty-five or thirty, some will walk into a crowded room, into weddings or graduations or wakes, and for them, he's no longer there. Totally disappeared. Or perhaps they never saw him in the first place. And he doesn't seek them out. They're not his people.

But you? You've come to count on him, this guy in black. And as the years pass, he starts showing up on a daily basis. Booze has your back.

In fact, he knows where you live. Need some energy? Need some sleep? Need some nerve? Booze will lend a hand. You start counting on him to get you out of every fix. Overworked, overstressed, over-whelmed? Lonely? Heartsick? Booze is there when you need him most.

And when you don't. Suddenly, you realize booze has moved in. He's in your kitchen. He's in your bedroom. He's at your dinner table, taking up two spaces, crowding out your loved ones. Before you know it, he starts waking you up in the middle of the night, booting you in the gut at a quarter to four. You have friends over and he causes a scene. He starts showing you who's boss. Booze is now calling the shots.

You decide you've had enough. You ask him to leave. He refuses. A deal is a deal, he says. He wants payback and he wants it now. In fact, he wants it all: room and board, all your money, your assets, your family—plus a lot of love on the side. Unconditional love.

You do the only thing you can think to do: you kick him out, change the locks, get an unlisted number. But on Friday night, he sneaks back in, through the side door. You toss him out again. He's back the very next day.

Now you're scared. This is the toughest thing you've ever dealt with. You decide to try the geographical cure: you quit your job, pull up stakes, relocate to a new city where no one knows you. You'll start afresh.

But within days, booze comes calling in the middle of the night. Like all loan sharks, he's one step ahead of you and he means business.

This is how it happens.

This is addiction.

———————

PART ONE

Singing Backwards

1.

The Monkey Diary

THE BEGINNING OF THE END

To be rooted is perhaps the most important and the least recognized need of the human soul.

—SIMONE WEIL

For me, it happened this way: I took a geographic cure to fix what I thought was wrong with my life, and the cure failed.

Much later, I would learn the truth: geographic cures always fail, especially when they're designed to correct problematic drinking.

Of course, that wasn't how I saw it at the time. In the winter of 2006, when I pulled up stakes and moved to Montreal, I was full of hope. Hope that my fabulous new career would blossom. Hope that my long-distance sweetheart and I would flourish in this new city. Sitting by candlelight at my farewell dinner, these were the dreams I shared with my closest friends.

The third hope I kept to myself: that with this move, my increasingly troubling drinking habits would miraculously disappear. That my nightly craving for a glass of wine—or three—would go *poof.*

I was full of new resolve. I had made a New Year's resolution never to drink alone. I had made that promise to my sweetheart, and I intended to keep my word.

It was an icy blue February afternoon when I first dragged my suitcase up the marble stairs of Sam Bronfman's faux castle on Montreal's Peel Street, a Disneyesque confection that had been headquarters to the world's largest distillery for many decades. Donated to McGill University in 2002, Seagram House had taken on new life as Martlet House, named for the small red bird on the university's crest, believed to be blessed—or was it cursed?—with constant flight.

A martlet never rests. I chose to see this as a happy omen. I was looking for signs that I had made the right decision in accepting the big job of vice principal of McGill, in charge of development, alumni, and university relations. I had left my beloved home in Toronto and a successful career in journalism. I took this Martlet business seriously.

As vice principal, I was ushered into Bronfman's large second-floor office, the very same place where the booze baron had hosted Joe Kennedy and Al Capone during Prohibition—or so the story goes. It was here that I would sit, at his massive hand-carved desk, ensconced at one end of an airless chamber, walled with recessed curved bookcases and ornate oak paneling. The history was impressive. Once upon a time, the office had been, too. But when I arrived, stained green carpet, broken overhead fixtures, and the lack of natural light made the room oppressive. Still, it had loads of potential. I was optimistic.

In honor of my arrival, a fellow vice principal had placed hot pink gerbera daisies in a jaunty citrine vase. There were welcoming bouquets from the principal and others, and a vast array of notes and cards—a happy distraction on my first day. My gut was speaking to me, but I chose not to listen.

Over time, I grew to dread that behemoth of a desk, and all it represented. But on the first day, its novelty was a distraction. My effervescent blond assistant, only two years out of university herself, perched opposite me, pulling out the secretary's table to write on. She introduced me to a fat binder and handed over a pile of documents for my signature. Most of all, she was interested in securing a date for my welcoming reception. Her top choice was St. Patrick's Day—or St. Patio Day, as she liked to call it, the booziest day on the Montreal calendar, and her personal favorite. (She was single and anxious to change that status.)

Five weeks later, she made it happen: the majority of my new staff—there were more than 180 in all—crowded into the ground-floor boardroom of Martlet House for coffee and croissants as the principal welcomed me to McGill. I was in charge of mobilizing this group to launch the largest campaign in the university's history, a $500 million fund-raising effort that would change the face of Mc-Gill, boost research, help students. The principal was a woman I deeply admired. My heart was full. My geographical cure was going to work.

For the first months, I spent many nights behind that big Bronfman desk. Sometime around six in the evening, as the last of my staff headed home to husbands and wives, children and friends, I would walk half a block to the small café on the corner, order a takeout salad, and chat to the owner in broken French, getting ready for another evening at work. Occasionally, I'd stay past midnight, and return on the weekend. I was used to long hours. I had no real friends in the city and my learning curve was steep. The previous vice principal, recruited from Stanford, had left before her tenure was up. Most credited her with professionalizing the fund-raising machine of McGill, and it was my job to continue the process.

I dug in hard. Senate documents, issues of governance, fat background packages on donors: these were the easy files. What was confounding was the management challenge, picking up where the Stanford woman had left off. At bedtime, I'd close the day with a few emails, place my BlackBerry on the pillow beside me, and struggle through a few pages of *The First 90 Days: Critical Success Strategies for New Leaders at All Levels*, a farewell gift from a seasoned manager back home.

Shutting off the light, I'd review my day in terms of the "monkey rule," advice I'd received from a renowned university president. "There is only one way you can fail at your new job," he had warned. "Your key reports will come into your office with monkeys on their shoulders. When they leave, make sure their monkeys are on *their* shoulders, not yours." Great advice; tough to follow. I'd fall asleep, visions of monkeys dancing in my head.

By spring, the light of Montreal was transformed. Patio season had arrived, and my assistant's agenda was full. Each morning, she'd bring me a fresh installment of romance along with my coffee and documents. As she rushed out each evening, glowing with possibility, I would crank open the latches of the leaded glass windows behind the Bronfman desk and let the sounds waft up from the back alley. The popular Peel Pub, a rowdy favorite with undergrads and their out-of-town visitors, was only doors away. So too was Alexandre, a cozy local. In the morning, my assistant would frown at the open windows: "Why on earth would you want to look at a brick wall?" How could I explain that I found the nighttime sounds strangely comforting? The staccato chatter of busboys on their smoking breaks, hauling buckets of empties to gray bins, grabbing a quick smoke before they headed back to work; the occasional burst of laughter; furtive snatches of a melody, a bit of bass.

It reminded me of what a friend once said of sex on antidepressants: "I can manage an orgasm, but it seems to be happening to someone down the hall." Life, once removed.

My sweetheart Jake—a writer living thousands of miles away— had just proposed to me. One week before my move, we had escaped to a remote island in the Bahamas. There, on a deserted beach at sunset, he had asked me to marry him. I had said yes.

Over the years, Jake and I had had many honeymoons. For a decade, we had spent as much time together as possible, summering at Jake's octagonal wooden houseboat in the wilds of northern Ontario, wintering in each other's homes. In between, we traveled: Paris, London, Mexico. We each had one child: a daughter, Caitlin, for him; a son, Nicholas, for me. Born six months apart, they had been eleven when we met. We had raised them with dedication and delight, in tandem with our former partners, and each other.

For years, it had been a perfect arrangement. In summer, we moored by pink-streaked granite and woke to otters stealing from our minnow bucket; on our morning swims, there were occasional moose or bear sightings. At night, nursing glasses of Irish whiskey, we would sit under the stars on a handmade driftwood bench, our personal playlist wafting across the water. In winter, we read and stoked the fire and wore flannel fish pajamas while we cozied up to watch classic movies. "We have something better than everyone else," Jake would say, and I was certain he was right. In Jake's presence, I felt like Grace Kelly in *Rear Window*—the cosmopolitan girl, head over heels in love with the globe-trotting Jimmy Stewart. More often than not, it was bliss.

For the first two months in Montreal, I was buffered from the full-frontal blow of my decision to move, living two minutes from Martlet House in an executive apartment hotel. In many ways, it felt like an extended business trip. Jake—whose nickname was Jackrabbit—had shipped a package to the front desk for Valentine's Day: I am sure I was the only person in that hotel with a stuffed jackrabbit on her pillow.

Each night, Jake would tell me about his writing day, a world I understood intimately. "Feels like cracking concrete with my forehead," he'd say. "Tell me about your day, baby." Holding on to that rabbit, I'd try to entertain him with the complexities of my new world. I'd always end the same way: "Looks like someone forgot to book my return flight," I'd joke. Neither of us ever laughed. There was a peacefulness to our nightly calls. He had just had an unexpected hip replacement, and I had flown out to nurse him. He was anxious to heal, to come to Montreal, to take a crack at the city. I was keen to have him by my side: I was growing more lonely by the day.

By June, I'd stopped spending evenings in the stuffiness of my office. Night after night, I'd lug my work to the warm glow of Alexandre, settling in at the same cozy table with my BlackBerry and my reading. I could see other people, and it eased the deep sense of isolation. Night after night, the same waiter would bring me the first of three glasses of crisp Sauvignon Blanc, a warm chèvre salad, and a baguette. Every evening, he did his best to change my habits. "Escargots, madame?" "Non, François." "Steak tartare?" "Non, merci, François. Un autre verre."

With that first sip, my shoulders seemed to unhitch from my earlobes. With the second, I could exhale. I loved the way the wine worked on my innards. That first glass would melt some glacial layer of tension, a barrier between me and the world. Somehow with the second glass, the tectonic plates of my psyche would shift, and I'd be more at ease. Jake used to say it this way: "When you drink, that piano on your back seems to disappear."

I had always taken my work seriously—maybe too seriously. Somewhere between the first and second glass, I'd take a fresh look at a problem, or find an answer to some complex question. Suddenly, it all looked simple. By the third glass? Well, that one just took my clarity down a notch, and I'd know it was time to go home.

Did my evenings at Alexandre count as drinking alone? I tried to fudge it in my mind, thinking of it this way: I wasn't *exactly* alone.

There were people at neighboring tables. Besides, I had no one to have dinner with. But in my heart, I knew the truth: I was breaking my promise—not only to Jake, but to myself. I was drinking because I was lonely. I was drinking because I was anxious. This wasn't Grace Kelly pouring a glass of Montrachet for Jimmy Stewart. This was something else, something I had never encountered, and it felt wrong.

I don't know what month I began picking up another bottle on my way home, in case I wanted a glass before bed. But I do know exactly when I began sleeping through two alarms.

One fateful June evening, after a particularly difficult interchange with a senior employee, I headed off to Alexandre. Here is my journal:

Four. I had four last night. Maybe it was five. One was vodka. And I slept through both alarms.

My boss' car left for the country and the annual executive retreat, and I missed the ride. The car came to pick me up, with her in it, and found no-one waiting. I will have to resign. In 30 years of professional life, I have never made an error like this one.

I made it by 11:00, but there is no mending what is broken.

My boss asked me: "Did you take a sleeping pill last night?"

"No," I said.

"I was hoping you'd say you had."

Two weeks later, in one of our regular meetings, she asked me how I was doing. I surprised myself with the answer: "I don't know how to explain it, but I am losing my voice." And somehow, this was true. I was losing myself in Montreal. And missing journalism—my writing, my world—was only part of the story.

That summer, Jake bought me a beautiful gold engagement ring, hand-carved with delicate leaves—or were they bird feathers? Either way, it spoke to our love of nature, and of our time in the woods together. I stole two weeks at the houseboat. We swam each morning before breakfast, and indulged in our long morning meals, sitting on the driftwood bench, the table laden with fruit and eggs and coffee. It always ended the same way: "Come here, baby, sit on my lap and I'll rub your back." Jake and I would kiss before we parted for our morning chores. By mid-afternoon, he would be baiting my hook, mid-river, the two of us on one of our four-hour adventures in either the Boston Whaler or the classic wooden boat.

But this summer, the talk was less about writing and more about BlackBerry reception. I might be on vacation, but the monkeys weren't taking a holiday. Rumor had it that the principal's husband had tossed her BlackBerry in the lake one summer, so frustrated was he by her constant emailing. I thought the story was apocryphal: it was hard to imagine her unrufflable husband—screenwriter and yoga master— tossing anything. Still, Jake found the story amusing, especially when my own work habits tested his patience. His mother was concerned: "You look exhausted," she said to me. "Something about this job isn't good for you." I held her hand and told her it would be fine.

By fall, my loneliness was overwhelming. But like the unhappy couple who decides to have a baby to fix their marriage, I had started to work with a real estate agent to purchase a home. In the meantime, I moved into temporary digs on the executive floor of the new student residence. My peripatetic ways were raising alarm bells with the principal, and so they should have been. Most weekends, I was flying home to Toronto, BlackBerry in hand. During the week, I'd troop through a selection of condos and houses, rejecting them all. They looked like movie sets to me, backdrops for a life that had nothing to do with mine.

The frosh had arrived. Each night, gangs of fresh-faced kids would pour out of the residence, eager to down another heady gulp of Montreal nightlife. From where I sat, they seemed to have the city on a string. Me? I was up on the fifteenth floor, with a glass of white wine, checking out real estate listings, lost as lost could be. I had a big job, a life partner halfway across the country, and not a true friend in sight. My summer holiday with Jake was long over and I felt like my life was close to over as well.

All that fall, the residence rocked late into the night. Sometimes, all night. "Jumpin' Jack Flash" pulsing at 2 a.m. The gravelly voice of Leonard Cohen trailing down halls. Four years earlier, my own son had headed off to university himself, taking his guitar but leaving a CD on my pillow, with a note: "If you get lonely, play this music LOUD."

This residence felt as close to home as it was ever going to get in Montreal. I liked wandering the corridors, listening to the Korean student play the grand piano in the foyer, watching young girls in bunny slippers giggle over pizza. One evening, when I was coming home late, the elevators opened to reveal three semi-nude guys, all dyed various shades of red, with matching towels tied around their waists, their heads encased in Molson Canadian boxes, with eye slits.

"Well, hello, miss! I take it you're new in town?"

All three were weaving slightly.

"Not as new as you," I said. "I'm one of the vice principals."

One head case straightened up.

"Oh, sorry, ma'am!"

He wiped his hand on his towel, and gave my hand a good pumping. "Nice to meet you!"

American, I thought. From the South.

"Nice to meet you, too," I said as they drifted off into the night. The elevator doors closed. I thought: "I'm the oldest coed in this place."

As midterms got closer, the music got a little softer, but the drinking never seemed to slow down. Girls sobbing in the front lobby, their eyes smudged black with mascara. Guys lying facedown on the sidewalk, passed out, their pals swigging beer beside them, texting. Once in a while, the elevators would smell of vomit.

My life was lonely beyond measure. There was the occasional visit from an out-of-town friend or a McGill parent in town for graduation, or someone checking on a troubled son or daughter. Once in a while, I would have a meal with Professor Dan Levitin, musician and producer turned neuroscientist, author of *This Is Your Brain on Music*. Dan lived alone with his dog Shadow. I liked hearing about his new pal Sting, his old pal Joni Mitchell, Rosanne Cash, Tom Waits. He was a moderate drinker, a lover of puns, and had great taste in restaurants. He was also single. After a while I felt awkward seeing him. With regret, I let our friendship wane.

One night before Christmas, François came up to me, looking concerned. "Madame, I think you are very, very lonely. I think you are the most lonely woman in the world."

"No, François, I am not."

François looked unconvinced.

"I am just very busy." I picked up the pile of papers on the banquette.

"Oui, madame."

The geographic cure was not working. I knew it, and others were beginning to suspect it as well. That New Year's, Jake and I wrote out our resolutions for each other, as we always did, signing one another's promises. This year he looked up from his own list and interjected as I wrote mine: "No more than two drinks on any one occasion," he said. "And no drinking alone." "Don't you think three is more realistic if it's an evening out?" I bargained. "Three over three hours," said Jake. He didn't look convinced. And so I wrote: "Given the genetic predisposition to alcoholism in our family, I do resolve to do the following: to limit my drinking to two drinks in social situations, three over three hours; no drinking alone, ever; nine drinks total a week. If I have broken any of these rules within six months, I promise to get help." Jake and I signed each other's sheets, and dated them: January 1, 2007.

Jake wasn't the only one worried about my drinking. My son had noticed a big change, and was vocal about it. My sister was quiet, but I could read her silence. Our mother had had a serious drinking problem. Me? I was beyond worried. I decided to take action: I called an addiction doctor, and booked his earliest appointment. Sadly, it was March.

Most of all, I wanted to go home. This was not an option, or I didn't see it as one. At Martlet House, we had closed a very successful year: a record year of fund-raising. I was proud of my association with McGill and with this achievement. In two weeks I was taking possession of a beautiful light-filled condo in an historic building. In nine months, the major fund-raising campaign was going public. I was in the middle of helping to recruit a cochair for the campaign. I was on deadline and I took it seriously.

So, I did the only thing I could think to do: I started a drinking diary. My sister suggested rewarding good behavior with stickers. I ducked into a toy store and bought the first ones that jumped out at me: monkeys. Perfect. I would get this damn monkey off my back.

Of course, as I learned much later, this is how the ending always starts.

You know you're drinking too much, so you decide to keep a tally. And if you're like most, you keep this tally hidden. In your wallet, or your underwear drawer. Last night you drank four. Or was it five? Tonight, for sure, you will do better.

This is how it begins. You set some rules.

Maybe you switch from red to white (less staining on the teeth).

Or maybe it's no wine; only beer.

No brown liquids, only clear. (Vodka doesn't smell, does it?)

Only on weekends.

Never on Sundays.

Never, ever alone.

The problem is: The rules continue to change. Your drinking doesn't.

You take up running or swimming. (In my case, it was power-walking. People who power-walk can't be alcoholics, can they?)

You start to wake at four in the morning. (Doesn't everyone wake at four in the morning?)

You promise to do better tonight, to drink less.

Only you don't.

In fact, the only commitment you seem able to keep is the diary. It tells a story, and the story is starting to look scary.

Worse still? This is only the beginning of the end.

Like many a drinking diary, mine started off well. For a few days, the monkey stickers began to accumulate: I had kept to my limits. Of course, I kept the diary hidden. (What vice principal pastes monkey stickers into a journal?) But it wasn't long before those stickers petered out. Alcohol is a formidable enemy: once you name it, it digs in hard.

I said this to the addiction doctor in March. He nodded. "How do you feel about alcohol now?" he asked. "I love it." He frowned. "And I hate it." "Be careful," he warned. "Alcohol is a trickster. And using alcohol to cope is maladaptive behavior."

One spring evening, I had dinner with the eloquent dean of medicine, Rich Levin. He was newish to McGill, having moved with his wife from New York, and he had had a difficult day.

Rich was a martini drinker, and he ordered one, then another.

"Why did you come to Montreal, Rich?"

"I came here for the waters."

I fell for it. "The waters?"

"Turns out I was misinformed."

I looked puzzled.

"*Casablanca*."

"Another drink, Rich?"

"Never, my dear. You know what Dorothy Parker says."

The next time I saw him, Rich pulled a gently used cocktail napkin from his pocket and handed it to me. There were Parker's words, emblazoned beside a martini glass: "*I love a martini—but two at the most. Three I'm under the table, four, I'm under the host.*"

That night, I pasted the napkin into my diary. Beside it I wrote: "*I am bullied by alcohol. I am hiding behind it.*" I knew the jig was up.

Days later, on Father's Day morning, I learn that my cousin Doug—childhood confidant and best friend—had been killed by a drunk driver, on his way home from his mother's eightieth-birthday celebration. His young daughter, the youngest of four, was in the front seat. She survived but was severely injured.

It was a sunny Sunday morning, and I remember thinking: "What else do you have to lose to alcohol before you give up?" I had already lost a big part of my childhood, now my cousin—and I was losing myself.

I pulled out a bulletin board and tacked a piece of paper with four handwritten words at the top: "The Wall of Why." As in, why I needed to give up drinking. Or: why I needed to avoid dying. The diary was no longer working. In fact, it had never worked. For the first time, I was terrified this habit might kill me.

I spent an hour filling the board with images and words I loved. In that condo, I had very few photographs—one of Nicholas with his arm around me, after winning bronze at a rowing regatta; one of Jake casting a line off the houseboat deck; one of my dog Bo. There were so many faces missing. I took out my fountain pen and wrote the names of others on pieces of white paper, pinning them carefully to the board. Then, I added several pieces of prose—Annie Dillard, Simone Weil—and some poetry: "Love after Love," by Derek Walcott.

Then I got down on my knees and said the only prayer I believed in, words from T. S. Eliot's "East Coker":

I said to my soul, be still, and wait without hope
For hope would be hope for the wrong thing; wait without love,
For love would be love of the wrong thing; there is yet faith.
But the faith, and the love, and the hope are all in the waiting.
Wait without thought, for you are not ready for thought:
So the darkness shall be the light, and the stillness the dancing.

Within weeks, Jake and I would find our way to a recovery meeting in a church basement. He held my hand while tears rivered down my cheeks. For an hour I listened to a roomful of seemingly happy people share their stories, their faith, their gratitude. As they started to stack the chairs, a tall black stranger in a funky hat came up to comfort me. "Darlin'," he drawled, "believe me, whatever you did wrong, I did way, way worse."

Every season has its own soundtrack: that summer, it was Keith Jarrett's introspective *Köln Concert* wafting over pink-streaked granite, keeping us company as we drank cranberry juice and soda with our meals. Jake's precious mother had just died a difficult death. When Jarrett felt too haunting, Jake would toss in a little Frank or Van to keep the tone romantic. "I'm making love to you with my playlist," he'd call out from his computer, and I'd be enveloped, newly sober, in a fresh cocoon of sound.

But for the rest of the world, the summer of 2007 belonged to the defiant Amy Winehouse: *"They tried to make me go to rehab. I said No, no, no!"* An earworm if ever there was one. The point wasn't lost on me as I headed back to McGill, having tallied my first seventeen days of sobriety in the north woods of Ontario. Checking my BlackBerry as I cabbed in from the airport, I found myself humming along. *"No, no, no!"*

Little did I understand that it would be more than a year before I was able to secure any meaningful sobriety, to put alcohol somewhat solidly in my rearview mirror. It would be three years after that before I regained what could be called a true sense of equilibrium. And it would take all my journalistic skills to put what was killing me—and as it turns out, a growing number of women—into some profound and meaningful context.

In the meantime, I was about to lose many things I cared about: my livelihood, my heart, my gusto. And before things got better, they were going to get as tough as tough could be.

2.

Out of Africa

A FAMILY UNRAVELS

One always learns one's mystery at
the price of one's innocence.
—ROBERTSON DAVIES

I had a bifurcated childhood, split perfectly down the middle between joy and distress. Most of the latter was alcohol-fueled. My sister and brother will attest to this, and my mother will as well: there was great happiness, despite the extended absences of my peripatetic father, followed by years of terrible despair, years we barely survived.

What we don't agree on is when it all changed. For me, it split pretty tidily this way: before South Africa—a move we made when I was nine—and after South Africa. South Africa was the hinge experience. Once we had been there, it seemed there was no turning back.

Before we moved, there were many memories, but none so dominant as my mother's devotion to her parents. Night after night, I fell asleep to the sound of her typewriter keys as she wrote her long letters home. Handel or Beethoven on the record player, clackety-clack. Telling them of her life in a small northern mining town, with three small children, where the whistle blew every evening to signal that the miners' day had ended. Clackety-clack. Writing of life alone with those small children. My father in Africa or Australia, a geophysicist overseeing exploration in the outback. Clackety-clack. Once in a while she would go to her bridge club. Kissing me when she returned, she smelled of cold air and clean hair and Guerlain's l'Heure Bleue. But those evenings were rare. Most evenings, I fell asleep to the comforting sound of her keys.

And then glorious silence: come June, the typing would stop and we'd hit the road.

Year after sunburned year—long before people worried about global warming or SPF—we would escape for the entire season. As soon as school was out, my mother would load up the car and head off down the highway. In the trunk would be our tartan cooler, the car rug for picnics, plus an entire suitcase of library books. In the backseat: the dog, my sister, my brother, and I, unencumbered by care—or seat belts, for that matter.

On paper, my mother would say we were Protestants. But in reality, heading to the cottage was our religion: we were the true believers. Not that we worshipped in just one spot. As newlyweds, my parents had honeymooned at my father's family place, a log cabin on a sheltered teacup of a lake near Algonquin Park, the same lake where iconic Canadian painter Tom Thomson planned to honeymoon before he mysteriously drowned. But after that initial trip, they split their vacation time between their families' summer homes. And since my father's holiday time was limited, more often than not we would find ourselves nestled in the bunk beds of my mother's childhood cottage on a stretch of Georgian Bay, a place where August storms swaggered in at night, tossing sailboats at their moorings, working their bonsai magic on the pines.

Thanks to my two grandfathers—both of whom had fought in the First World War, one as a fighter pilot, the other having his leg shattered at Passchendaele—there were two log cabins we called home. During the 1930s, they and their spunky wives had searched the north country for land, tenting with their children before the cottages were built. In my maternal grandparents' case, they bought a local farmer's log home for five hundred dollars in 1930 and had the thick hemlock timbers numbered and transported by horse and wagon to be reassembled by the shores of Georgian Bay. My paternal grandparents, on the other hand, built a tidy one-room log place from scratch, adding little pine bunkhouses along the shoreline as their family grew.

As a result, we gorged on summer in two distinctly different places. At the little lake cabin, we would fall asleep to the mournful call of loons, snug under heavy red Hudson Bay blankets, in flannel pajamas my mother had warmed by the fire, our hair smelling of wood smoke. My sister Cate and I would whisper by the dying light of the woodstove. What was that noise? Was it a bear? Or a ghost? I was sure there were ghosts. Poor Tom Thomson, vengeful in his soggy plaid shirt, rising from his watery grave to return to his never-to-be honeymoon spot, wielding an axe. Always an axe, to give us forty whacks.

Before I knew it, morning would break with a *slam*, my grandmother's screen door announcing she was up, coffee on, porridge started. Time for the morning paddle to the lodge to see if the paper had arrived. Within minutes we would be off, her voice ringing clear across the mirrored water: *"By the li-i-ight of the sil-ver-eee moo-oo-oon . . ."* Another day had begun, a day of snooping in the woods, racing to the raft, horsing around with the Patterson boys.

At the other cottage, days and nights were different. There I would fall asleep to the sulky rhythm of Georgian Bay and the tinkling sounds of masts, the sweet taste of marshmallow in my mouth and even sweeter comfort of my cousins. By day I'd wake to thick wedges of sunlight on the painted floorboards and the whirrrr-dee-dee-dee of the birds. In a flash, I'd be downstairs, joining Dougie as he cracked open a new variety pack of little boxed cereals, dousing his bowl of Frosted Flakes in chocolate milk because *shhh*, the mothers were still sleeping. Off we would tramp in our still-damp bathing suits to our secret fort at the Point. Back to the cottage to head out in the *Swallow*, our bathtub of a homemade rowboat. Adventure after adventure, punctuated only by meals, served by my mother and aunts and grandmother on little birch-bark place mats, ones sold by the "Indians," said my mother, "when they used to tent on the Point."

All week long the cottage was a women-and-children affair. But on Friday afternoons the air would begin to crackle. For hours we would line ourselves along the top of the split-rail fence, chirpy faces trained toward the curve, looking for the first sign of a Buick. My mother would head into the bedroom to brush her freshly washed hair, put on lipstick, and emerge transformed: burnished and blond for my dad. I thought she looked like a movie star.

For the next two days there would be laughter: games of charades, rounds of bridge, impromptu skits. Tall shoulders to be tossed from, into the water; strong arms to help us build boats and forts. Handsome men drinking "Hey Mabel, Black Label" beer after splitting logs and stacking the woodpile. Was there too much drinking? I have no idea. All I remember is that most of the adults smoked cigarettes or pipes—and we did, too, sneaking them into our homemade tepee. It was a poor plan. The smoke billowed out the top and we were caught, red-handed, forced to chain-smoke until we turned green.

At night, lying under white sheets, little needles of sunburn prickling our shoulders, our noses peeling for the umpteenth time, my cousin and I would decide that *no*, we weren't going to sleep, not when the adults were telling dirty jokes downstairs. And so we would eavesdrop, and then whisper very quietly, because "*for the last time, kids,*" my uncle had warned, "*it's time to go to sleep!*"

Then it would be Sunday night, and we would all wave as the cars, honking, disappeared around the curve, and the cycle would begin again.

Often there was a visitor I loved: the painter A. Y. Jackson, a close friend of my grandparents. A bachelor with an infamous appetite for my grandmother's jam—jam that would dribble down his sweater vests along with his cigarette ashes when he chuckled at my grandfather's jokes. Looking at his belly, I knew why Aunt Esther had never married him. "I'll be away many weeks of the year," he had warned when he proposed. "Make it fifty-two, and I might agree," was her response.

Or that's how the story went. Maybe he never really wanted to get married. Maybe he was afraid of marriage the same way he was afraid of fire. In one of the cottage bedrooms, he had had my grandfather install a thick rope, attached at the windowsill so he could shimmy down it in the event of a blaze. (He never used it, but Dougie did, when we played hide-and-seek.)

Clearly, this was a man who liked to escape, just like my father. He stole my heart because he taught me how to steer a paintbrush with my thumb, and because he painted a naughty little sketch of a Shell station with the *S* missing. He was just about the only bachelor I had ever met, a rambling guy whose snowshoes hung on the wall beside the fireplace. But I used to think that maybe he'd outfoxed himself, taking all those painting trips and somehow forgetting to get married and have his own little family to go home to.

Year after sunburned year, this was how we lived. If my father was away more than most—and he always was—summers buffered my mother from her pervasive loneliness. She flourished near family, and so did we.

When this chapter ended—when her parents died too young, and her drinking started—I used to think that those summers at the cottage were like money in the bank or gas in the tank: she had accumulated so many good memories for us that it took a long time to get to zero.

Of course, we did get to zero, and far, far beyond in those many years when her drinking was dire, when she seemed to give up eating and sleeping, weaving down the halls like a passenger on a bumpy train, the sound of ice cubes declaring her arrival. But that was much later, after the cottages had been split up and the cousins had scattered, a long time after Africa.

Rural South Africa, 1963

The birth of anxiety.

An outdoor music festival. I am ten years old, about to sing my first solo, "All Through the Night."

Suddenly, the large Afrikaans woman next to me starts to slap my arm. I can't understand a word she's saying. (Turns out, I'm not clapping for her son, my competition, who's taking a bow.)

Now it's my turn to take the stage. I freeze.

The audience starts to mumble.

The judge walks over to see what's wrong. My father comes, too. He talks me into performing. Maybe after the lunch break, suggests the judge. I nod. I can do this, on one condition. I don't think I can face the crowd. The judge smiles. No, it's not essential to face forward.

After lunch, I keep my promise. On uncertain legs, I climb the stage. I turn to face my teacher, Mrs. Duplessis, at the piano. My back is to the audience.

At the end of the afternoon, I'm awarded second prize: first for voice, marks off for delivery.

Many years later, as I wrestled with major depression and a serious case of writer's block, the judge's verdict would become a constant in my head: "First for voice, marks off for delivery." Code for: you're failing.

For decades I forgot this incident. Then it reappeared, just as I was to deliver a cover story on teenage suicide. Frozen at my computer, I would hear the judge's words, over and over, in my head.

In Montreal, as my world unraveled on the fifteenth floor of the student residence, I heard it again: "First for voice, marks off for delivery."

What does this story have to do with my drinking?

Everything.

Liquor soothes. It calms anxiety. It numbs depression. Ask any serious drinker: if you want to find your off button, alcohol can seem like an excellent choice.

But not when you're ten.

Back then, as I sat with my parents on sticky chairs on an unforgiving African afternoon, my confidence was deeply shaken. I was pink with humiliation. And I felt my parents' confusion at my behavior. I had always been top of my class. I had accelerated at school. I had never blown anything this badly before.

It was a wobbly time for me.

For my parents, our move to Africa was their great romantic adventure. For the first time in their marriage, they had finessed their situation and were truly together in a sustaining fashion: making a home on an acre in Mount Ayliff, a village of a hundred people nestled in the hilly terrain of the Transkei. All of a sudden there was a cook, a maid, a garden boy. Each night, my father would park his Land Rover out front and saunter through the door, darkly handsome in his khaki field clothes. He would light the propane lanterns—there was no electricity—and cast a warm glow through the long-halled house. My mother was delighted.

For me, he was our bachelor father, and everything seemed new.

At dinner, when my mother's back was turned, he'd line his peas on his knife and toss them down his gullet, pressing his finger to his lips: "Shhhh!" "What, John? What did you just do?" We'd giggle. When she wasn't looking, he'd open the fridge and swig milk straight from the bottle. This guy didn't seem to know the rules.

At bedtime, he'd read to us: Arthur Ransome's *Swallows and Amazons*, C. S. Lewis's *The Lion, the Witch, and the Wardrobe*. My mother was in love with him, and we were, too: this dad was different than the person I expected, but he made the house hum. My mother was chatty, extroverted, radiant. They entertained, and there was laughter. This was an era of cinch-waisted dresses and "sundowners."

I knew I was meant to be happy. But for me there was a deep sense of foreboding, a shadow I could not shake. And it went deeper than the obvious disappointment that I was no longer my mother's primary companion, the eldest with special privileges.

More often than not, I felt like the bad fairy at the birthday party. There was a deep, subterranean rumble I could feel, although I couldn't put my finger on it. Something was not right.

It started with our trips to the library, back in Canada, when we were busy getting shots and passports. While the librarian was loading my sister's arms with animal books—books full of lions and poisonous snakes—I was reading stories about violence in the Belgian Congo, murder. I was deeply skeptical about this trip, and my fears seemed justified. When our plane landed for refueling in the Congo, en route to Johannesburg, I thought that the cleaners were boarding to kill us: I ran to the washroom, burst in on a man shaving, and promptly threw up. When my father introduced us to the snakebite kit in the kitchen, my fears were confirmed: this was African Gothic.

I had always loved school. But in Mount Ayliff's barren two-room schoolhouse—twenty students in eight grades—I couldn't understand a word being said. This was Afrikaans immersion, and I was lost. For the first time in my life, I was bullied on a regular basis. While two large boys would pin me to the ground, another would hold an insect close to my face, tearing its legs off, yelling at me for speaking English. I didn't tell my parents: as far as I was concerned, there wasn't much point. They didn't know Afrikaans, either.

Of course, in a very short time, my sister and I learned Afrikaans and Xhosa, too, the Bantu language spoken by our servants. I found a defender at school, a much younger English boy named Nicky Hastie, so staunchly loyal that I later named my son after him. My sister and I adopted a pet frog, named him Sam, and carried him to school in a little cardboard suitcase. He lived under my desk, and I'd peek on him when Mrs. Duplessis had her back turned.

In other words, we adjusted. While my mother developed a close relationship with our cook, the beautiful Ivy, we learned that the maid, Evelina, had a vicious temper and hated children. When my mother was out, Evelina would threaten us with a hot iron, chasing us down the long halls of the house. On more than one occasion, she burned a hole in my sister's favorite blue dress.

One Saturday night, we paid her back. Left in her care, we disappeared into our bedroom wardrobe, leading her to believe we had run away. Screaming at the sight of our empty beds, she ran to the servants' quarters to fetch Ivy. By the time Ivy showed up, we were safe and sound. Ivy left, and we hopped back in the wardrobe. Later, Evelina would get even: when we headed back to Canada, we knew she was planning to chop the heads off our favorite bantam chickens to cook them for dinner. But by then our little pack of three was well established: John and Cate and I were tight as tight could be, and that fact would never change.

On Thursdays, Cate and I would wander by the local jail on our way home from piano lessons, passing hard candies through the fence to the neighbor's former cook. Rumor had it that she had killed a younger servant, whom she had caught sleeping with her boyfriend, the garden boy. Maybe it was the garden boy she killed. We weren't concerned: we loved her for the corn bread she had cooked, and for the hugs she gave us when we first arrived in the village. We liked her smile (although I used to imagine her washing the blood off the knife, after she stabbed the person; I thought she must be very brave).

There was a political undercurrent in the village. We knew we were the last whites to live in Mount Ayliff, that soon this would be the first homeland given back to the blacks. One Saturday a group decided to speed up the process: they would burn the whites out of town. They warned two families to leave: ours and the doctor's. My parents woke us in the middle of the night and told us to get our clothes on: we had to decide whether to evacuate or not. We ended up staying. The crisis was averted, but the anger was real.

At Christmas, my mother gave both Ivy and Evelina presents. And on Christmas night our family joined a handful of others, sitting in church with the black congregation. Soon after there was a visit from the police: presents, they said, were not a good idea. Nor were the mattresses in the servants' quarters. My feisty mother was unfazed.

On vacations we would visit the Indian Ocean or a game park. Truth be told, I always thought those trips were risky business: I had been chased by a herd of warthogs and was certain it was only a matter of time before we were killed or maimed. While others enjoyed the view, watching monkeys try to pry open the car, I usually had my eyes trained out the back window, checking for a marauding rhino.

On weekends we would head off in our big boat of a Mercedes and end up at the Stanfords' ranch, where my parents would ride horses into the mountains, coming back with stories of baboons and more. My mother always looked so gorgeous on a horse, her hair windswept, a girlish joy on her face. She loved the adventure, and I thought she was remarkable, going off as she did, facing baboons. Remarkable, and a little reckless.

While they were gone, we would play hide-and-seek with the Stanford girls, discuss what little we knew of the facts of life, and look after our baby brothers. I liked those weekends: it felt like the cottage, with my cousins. I finally felt at home.

And then it was over. Before we headed back to Canada, my father presented my mother with a beautiful double-diamond ring to celebrate their African honeymoon. For two solid months we meandered up through Africa, from Zanzibar to Kenya, on to Egypt and Greece, Italy, Switzerland, and more, traipsing hand in hand like the happy band that we were.

Years and years later, after we were all married, my parents moved back to Africa, spending six years in Botswana. They took a trip one Christmas, down through South Africa, to visit old friends, stopping in at Mount Ayliff on their way. Our house was now a magistrate's office; the neighbor's house was lined with broken beer bottles. The garden was long gone. My mother said she was sorry she ever saw our home that way: it broke her heart.

By then it was the 1980s, and everything had changed. My mother's years of heavy drinking had cast a terrible pall over our entire family. While loneliness, depression, and anxiety would take me down, something took her far, far further. For years, she—like so many other women—added Valium to the mix, and it diminished her. The woman we knew in Africa had disappeared, and in her place was someone full of rage, bitterness, and despair. Most of all, she was completely unpredictable. One minute our mother was present; the next she had transformed into Medusa with a tinkling glass.

I often look at photos of all of us on the trip home from Africa—pictures in Florence and Athens and Cairo—and wonder if anything had started to go wrong. Dad slim and tanned in his Ray-Bans, Mum just steps ahead in linen and pearls, and those Jackie Kennedy shades. Both look inordinately happy. Did Dad know what was about to happen? Did she? Had the drinking turned dangerous already?

I think not. If anything, the time in South Africa was too good: my mother never quite readjusted to her loneliness again, and she never forgave my father for leaving her behind. It was decades before they would have another big shared adventure. And in the intervening years, life was very tough.

Or that's how I see it. It's one way I have been able to make sense of the story, to love her through the madness. It's tough to parse addiction, even when you've succumbed to it yourself.

Most of all, I like to remember my parents the way they were in Rome. My father, scooping my mother in his arms, carrying her up the hotel stairs. Her head tossed back, laughing: "Oh, John!"

I never saw her laugh quite that way again. Once her parents died, both from cancer when they were barely into their seventies, and my father resumed his overseas travel, she took her comfort in hard liquor. For years she seemed to live in her nightgown, wandering the halls by night, cursing by day. "This isn't living, it's existing," she would announce, over and over, her eyes belligerent. "I've had it up to here"—gesturing to her neck—"and the rest is toilet paper!"

At sixteen, I began my feeble attempts to leave home. If my father was around to hold the fort, I would pack a small bag and head off on foot, aiming for my cousins' house. I would never make it very far before my father would retrieve me, slowing down beside me in the family car. "Get in, Ann. Your mother needs you." And back I would go, in tears, to a house that felt like it was on fire, burning with rage.

Perhaps because he knew just how difficult my mother found his absences, and because he loved her, my father stood by her. We often wondered why, and how, he was able to do this.

Only once did he lose his temper in front of us, and the image is seared deeply in all our brains. He has lined us up in the kitchen—my mother, my sister, my brother, and me—making sure we are watching as he smashes all the bottles he has found in the house, breaking them, one by one, over the kitchen faucet. Bottle after bottle in his powerful hands, crashing on that slim bit of curved aluminum, until he punctures it and it begins to spout like a whale. Broken glass, spouting water. And we all stand, dumbfounded, tears of fury and despair rolling down my poor father's face, tears of contrition pouring down my poor mother's cheeks, all of us trapped in the hell of a family cursed by addiction, with no escape pending.

For years she was on the phone, drinking and dialing. Sober, she rarely picked up. Drinking? No call was too difficult, including to the police. More than once I had a date interrupted by a cop. "Your mother needs to find you," they'd say. "There's a family emergency." And I'd roll my eyes. If there was one quality I hated most, it was her disinhibition.

The bills added up. Once, my father had the phone cut off, and there we were, having to explain to our friends that ours was a phoneless household. There was no end to the embarrassment.

Years later, when we had all left home and wanted to visit them, my father would rarely warn us if she was on a bender. We'd just arrive, and be back in living hell. Later he learned to give us a tip-off. "Your mother is not feeling very well today," he would say. It was a feeble bit of code: I used to be angry that he didn't name it for what it was. But it seemed he could not.

Nothing changed, not for decades. She missed coming to the hospital when my son was born. She missed most important occasions. There were so many sad birthdays, depressing Christmases.

And then it ended. Sometime in her seventies, my mother followed the discipline of a weight-loss regimen and changed her drinking habits in the process. Over time my real mother reappeared, a tinier, softer version of her younger self, a woman who could manage to have two drinks of wine and put the cork in it, heading off to bed at ten in a way she hadn't in decades. It was deeply confounding, albeit welcome. There she is, at my son's graduation, beaming into the camera. Could this not have happened a few decades earlier? Saved us years of sorrow, fury, and pain?

"I don't touch hard liquor," she now declares. "I only drink wine."

And but for the rarest occasion, she's right: her drink of choice is white wine and Diet Coke. A downer and an upper, I always think. It's a curious mix, but who am I to argue?

Never underestimate real life. Things you think will never happen will occur, and more. Bad, and good. I know this, in my bones.

My gifted father, my precious sober parent, followed my mother on the same terrible path into alcoholism. His journey was very different—discreet, private, late—but it was alcoholic all the same. It took him down, and it broke everyone's heart. Most of all, it shook my mother. "He needs help," she would say. "I, of all people, know just how awful this can be."

But my father was a quiet man who grew more so with age: he wasn't going to reach out for help. When I told him I had started going to an abstinence-based support group, he took a long draw on his cigarette and said simply: "I went once. You know, I should have gone back."

A river runs through our family, through our bloodlines. It curdles our reason, muddles our thinking, seduces us by numbing all pain.

Sons and daughters, nieces and nephews: they all need to be vigilant. Tom McGuane once called alcoholism the black lung disease of writers. But I can't blame my profession.

Over the years, my mother watched me develop into a heavy drinker and she was concerned. One night, near the end of my bingeing days, I passed out in the bathroom at the cottage. I was sitting, fully clothed, on the toilet, pants up. She banged on the door.

"Did you know you were asleep in there?" she asked, incredulously, the next morning. "I never want you to go through the same hell I went through." I was taken aback. This was my straight-talking mother. I promised to slow down. Within a week I was sitting in a church basement. In truth, this is the one blessing her drinking gave me: it terrified me into quitting faster than I might have otherwise. Luckily, I quit when I was in the middle stage of the disease: there are many embarrassments I incurred, but many tragedies I avoided.

Over time, my mother and I began to see more of each other. When my father died, I insisted she get a passport and let me take her to California. "You're a great traveler, Mum," I said, watching her peer over the edge of a gondola as we headed up the rock face of a mountain. "You sound surprised," she said, smiling. "Why wouldn't I be? We were so lucky, you know, having that trip to Africa." Fifty years later, and it's still the highlight of her life.

The next morning, standing by the kitchen sink, she turned to me and said: "I will never be able to thank you enough for bringing me here."

"I love your company, Mum," I said, softly.

And I realized: it's totally true. I love my mother's company. I still love the way she puts on her lipstick at night and combs her hair. I love the way she looks like a teenager when we take out the Scrabble board. Most of all, I love her appreciation of my son's journey as an artist. She has an open mind, a generous heart, and an endless appetite for adventure.

She looked serious. "You know, I am heartsick when I think what my drinking did to all of you."

It happened like that, just out of the blue: the apology I had waited for, for so many heartbroken years. All I could think to do was hold her tiny frame close for a long, long time. She smelled good: of Guerlain's l'Heure Bleue, just like she always used to in the early days, when I was little. She was silent, and so was I.

I realized that I had forgiven her, as my son has forgiven me.

But our reconciliation only deepened my growing obsession: What was this thing that had taken us both down, albeit to such different levels and for vastly different lengths of time? What *was* this trapdoor that we both disappeared into? Down the bunny hole we both fell, into a seductive altered reality.

Why do some disappear for a few years, and others lose themselves for decades, or forever?

A river may run through my family, but it's also coursing through a significant and growing portion of femalehood. Ever so slowly, my search for answers, once so deeply personal, began to turn profoundly professional.

Sitting on a hard metal chair every Thursday night at my recovery group, I am surrounded by women of every age and every walk of life: young mothers with strollers rubbing shoulders with grandmothers; high school students with teachers, professors, musicians, dancers, actresses, life coaches, investment experts. Over by the coffee urn, tattooed beauties dating rock stars confide in well-groomed mothers of three. Rows and rows of women, banding together to find a solution to a problem both cunning and baffling.

Each Thursday, my home group welcomes newcomers. More often than not, they're female. More often than not, they're young: impossibly fresh-faced, if somewhat confused. Six months in, they're bringing their friends. A year in, they're starting to mentor fresh new arrivals. It goes on and on, the sisterhood of the newly sober.

"What's happening?" I always think to myself, nursing my tea in the second row of a capacity crowd, waiting for the meeting to start. "How on earth did we all get here?"

3.

You've Come the Wrong Way, Baby

CLOSING THE GENDER GAP ON RISKY DRINKING

One mojito, two mojitos, three mojitos … FLOOR!
—POPULAR BIRTHDAY CARD

Alcohol is ubiquitous in our society. It's hugely linked to our notions of celebration, sophistication, and well-being. It's how we relax, reward, escape—exhale.

Know your wines? You're affluent. Know your vodkas? You're hip. Know your coolers, your shots? You're young and female.

Alcohol abuse is rising in much of the developed world—and in many countries, female drinkers are driving that growth. This is global: the richer the country, the fewer abstainers and the smaller the gap between male and female consumption. The new reality: binge drinking is increasing among young adults—and women are largely responsible for this trend. What has not been fully documented, understood, or explored is that while women have gained equality in so many arenas, we have also begun to close the gender gap when it comes to alcohol abuse.

Women's buying power has been growing for decades, and our

decision-making authority has grown as well. The alcohol industry, well aware of this reality, is now battling for our downtime—and our brand loyalty. Wines with names like Girls' Night Out, MommyJuice, Mommy's Time Out, Cupcake, and yes, Happy Bitch; berry-flavored vodkas, Skinnygirl Vodka, mango coolers, Mike's Hard Lemonade: all are aimed at us.

When it comes to alcohol, we live in a culture of denial. With alcoholics representing just a tiny fraction of the population, it's the widespread normalization of heavier consumption that translates to serious trouble. In the Western world, the majority of us drink. And the top 20 percent of the heaviest drinkers consume roughly three-quarters of all alcohol sold. Episodic binge drinking by a large population of nondependent drinkers has a huge impact on society.

Most of us understand the major role that excessive alcohol use plays in family disruption, violence, and injury. Death? When compared to illicit drugs, there are many more deaths due to alcohol. According to Robert Brewer, leader of the alcohol program at the National Center for Chronic Disease Prevention and Health Promotion at the U.S. Centers for Disease Control and Prevention (CDC) in Atlanta, excessive drinking is the third leading preventable cause of death in the United States, after smoking and a combination of bad diet and inactivity. By conservative estimates, it's responsible for roughly 80,000 deaths each year: of those, 23,000 are female. Of the 23,000, more than half are related to binge drinking. For women, binge drinking is defined as four or more drinks on one occasion in the past month; for men, it's five.

According to a recent CDC *Vital Signs* report, female binge drinking is a serious, underrecognized problem: almost 14 million American girls and women binge drink an average of three times each month, typically consuming six drinks per bingeing episode. Meanwhile, one in five high school girls binge drinks. Among those who consume alcohol, the prevalence of those who binge drink rises from roughly 45 percent of those in their first year of high school to 62 percent of those in their senior year.

Women most likely to binge drink: those between the ages of 18 and 34 (in other words, those in their prime childbearing years), and those with higher household incomes. Binge drinking not only increases the risk of unintended pregnancies: if pregnant women binge, their babies are at risk of sudden infant death syndrome and fetal alcohol spectrum disorders. Meanwhile, for all women, binge drinking increases the risk of breast cancer, heart disease, and sexually transmitted diseases, among other health and social problems. "People who binge drink tend to do so frequently," says Brewer. "Most people who drink too much aren't addicted to alcohol. Most of these people are not dependent. What's the big picture? This is a major public health problem."

The United States is not alone in naming alcohol abuse a major health challenge. In Britain, Prime Minister David Cameron has declared binge drinking a national "scandal." Deaths from liver disease have risen 20 percent in a decade. Last year, Britain's chief medical officer, Dame Sally Davies, pronounced: "Our alcohol consumption is out of kilter with most of the civilized world." In a recent report by the Organisation for Economic Co-operation and Development, British girls were cited as the biggest teenage drinkers in the Western world: half of fifteen-year-olds said they had been drunk twice in the past year, as compared with 44 percent of British boys the same age.

Says Sir Ian Gilmore, past president of the Royal College of Physicians: "In the thirty years I have been a liver specialist, the striking difference is this: liver cirrhosis was a disease of elderly men—I have seen a girl as young as seventeen and women in their twenties with end-stage liver disease. Alcohol dependence is setting in when youngsters are still in their teens. This mirrors what we saw with tobacco, when women caught up with men on lung cancer."

If leaders in Britain are concerned, so too is much of the world. In 2010, the World Health Organization passed its landmark Global Alcohol Strategy, with 193 signatories. In the developed world, where noncommunicable diseases pose the greatest health threat, alcohol abuse is moving much higher on the health-risk agenda, and will continue to do so.

Is alcohol the new tobacco? In many ways, it is: a multibillion-dollar international industry dealing with market-friendly governments, enjoying a virtually unrestricted market for advertising, despite growing evidence that the substance has significant health risks.

In fact, recent research has revealed that alcoholism is a more serious risk for early mortality than smoking—and more than twice as deadly for women than men. German researchers found that compared with the general population, alcohol-dependent women were 4.6 times as likely to cut their lives short. The rate for men: 1.9 times higher than the general population. On average, both women and men died roughly twenty years earlier than those who were not dependent on alcohol.

"It is just like Virginia Slims," says David Jernigan, director of the Center on Alcohol Marketing and Youth (CAMY) at the Johns Hopkins Bloomberg School of Public Health. "Alcohol is a carcinogen, and it's particularly risky for women. Breast cancer is the poster child for that position. But the alcohol industry is selling young women on the notion that only really, really good things happen when there's alcohol. And to have really, really good things happen, you *have* to drink."

I came of age in the seventies, a heady time for women in North America. Smack-dab in the middle of second-wave feminism, my baby-boom peers and I headed off to university in our miniskirts and tie-dyed T-shirts, assured by Gloria Steinem and a host of others that the world was ours for the taking. We could, in Steinem's words, "grow up to be the men we wanted to marry."

Not for us the confining roles of our fifties mothers, harnessed to aprons, and what seemed like cookie-cutter lives. Not for us the quiet desperation, the Valium, the acquiescence.

And for me? Not the path of my mother. Sitting in my dorm room at Queen's University, unpacking my things—a brand-new copy of Joni Mitchell's *Blue*, a not-so-new edition of *A Room of One's Own*—I was unequivocal on one point: my life was going to look different. Very different. (Of course, it already did: I had rose-colored aviator glasses custom-made for this new chapter. I kid you not.)

If there was one trap I was determined I would never fall into, it was alcoholism. Risky drinking? Maybe. It was frosh week. There were keg parties and buckets of what we called Purple Jesus in my immediate future. I was five minutes from meeting my first serious boyfriend. Most conveniently, the legal drinking age had just been changed to eighteen, my age exactly.

But alcoholism? Never. Three times my family circled the residence, eager to get one last glimpse of me before they headed home. Not once did I look out the window. Not because I didn't love them, but because I did. Too much: I was deeply entwined in the family drama. I was ready to set out alone.

I was in good company: my whole generation wanted to start fresh. This was the school year of 1971–72. Politically, we were well steeped in the My Lai Massacre, just a heartbeat from Watergate. *Ramparts* was still alive and well on the newsstands, and the first issue of a new women's magazine was having its debut: it was called *Ms.*

Nothing, we were certain, would ever be the same. And frankly, nothing was—especially if you were female. Ours was the generation that would have it all: careers, families, freedom of expression, equal rights. Fulfillment on every level.

Did we have it all? With courage, endless creativity, and gusto, we certainly tried. Without a blueprint, many of us established excellent careers while raising children and nurturing marriages, juggling deadlines, child care, and housework. We experimented with full-time, part-time, flextime, and freelance work, nannies, day care, and shared babysitters, home offices, and virtual offices.

In many cases, our marriages were strained, and failed. Mine certainly did.

Could we have it all? Could we be the mothers we wanted to be *and* rise to the top? Many of us said yes—albeit sequentially. Or with enough help. Others said no, ditch the cape. The jury was out.

Today, more than thirty-five years after I graduated, women outstrip their male counterparts in postsecondary participation. "We Did It!" crowed a cover of the *Economist*, featuring Rosie the Riveter. "Women's economic empowerment is arguably the biggest social change of our times," trumpeted the article. An enormous revolution, with enormous ramifications. As the magazine warned, dealing with the social consequences of this victory will be one of the great challenges of the next fifty years.

More than forty years after Steinem helped launch a revolution, the debate rages on: can women have it all? These days, there are two powerful women at the microphone, offering a rich diversity of advice: Sheryl Sandberg, chief operating officer of Facebook and author of *Lean In: Women, Work, and the Will to Lead*; and Anne-Marie Slaughter, the first female director of policy planning in the U.S. State Department, a Princeton professor, and author of a persuasive *Atlantic* magazine cover story, "Why Women Still Can't Have It All."

In 2011, *Forbes* magazine called Sandberg the fifth most powerful woman in the world. For today's young working woman, Sandberg may indeed be the most powerful, period. Long before *Lean In* appeared in bookstores this year, millions had checked out her 2010

TED Talk, in which she offered women prescriptive advice on how to reach the C-suite. While calling today's women lucky, Sandberg cites the sorry news that women are not making it to the top of any profession anywhere in the world. Numbers say it all. Of the Fortune 500 companies, only twenty-one are led by women. Of 195 independent countries in the world, all but seventeen are led by men. Meanwhile, in the United States, two-thirds of married male senior managers have children, while only a third of their female counterparts can make the same claim.

While offering her prescriptions for change, Sandberg comes clean about some of the most difficult truths for working women. Top among these: while success and likability are positively correlated for men, the opposite is true for women. Saying a woman is "*very* ambitious is not a compliment in our culture," writes Sandberg. "Aggressive and hard-charging women violate unwritten rules about acceptable social conduct. Men are continually applauded for being ambitious and powerful and successful, but women who display these same traits often pay a social penalty. Female accomplishments come at a cost." Finally, she shares, "When reviewing a woman, the reviewer will often voice the concern, 'While she's really good at her job, she's just not as well liked by her peers.'"

I say: bless her for telling it like it is. She confronts and exposes some tough truths, among them: women need to smile more than men when negotiating for a raise. Smile, and continue to smile.

In fact, I think of her comments when I speak to Daisy Kling, a third-year Queen's student, currently on a transfer to Britain's Durham University. "Sexism is invisible, but it's real," says Kling. "Girls have more pressure on them to behave a certain way. You think you have the same rights as boys, so it's hard to understand why you feel held back. But there's a lot of pressure on girls to act ninety different ways at once: you have to be smart, you want people to take you seriously,

DRINK

you have to be attractive—but not *too* attractive, not slutty. You have to have experience, but not too much experience."

Sandberg's well-trademarked advice is aimed, in many ways, at Kling's generation. It amounts to this: lean into the boardroom table, not back; don't decide to "leave before you leave"—in other words, to opt out of the fast track before you've even had children; and make your partner a real partner. Her focus is what she calls the "Leadership Ambition Gap," and she's determined to help women eliminate the internal barriers that keep them from the corner office. All valuable, bracing stuff—especially for those about to embark on a professional journey.

Neither Sandberg nor Slaughter airbrushes the truth. As I write this, Slaughter's book has yet to be published. But I know from reading her *Atlantic* piece, and her *New York Times* review of *Lean In*, that her take and mine are aligned. This is a woman who admits to the complexities of long-distance parenting a troubled fourteen-year-old son. She confesses that "juggling high-level government work with the needs of two teenage boys was not possible." Says Slaughter: "Having it all, at least for me, depended almost entirely on what type of job I had." In other words, she believes there are times when you have to lean back. And while Slaughter's version of leaning back means trading one high-octane superstar position for an illustrious second, you have to love her candor.

As a young professional, I could have used both Sandberg's and Slaughter's advice. As I said, there was no blueprint back in 1977, when I started my career, two weeks after my wedding. When I gave birth in 1984, the term "second shift" had yet to be popularized. And when I proposed job-sharing to my editors at *Maclean's* magazine— job-sharing with the talented Canadian author, editor, and journalist Val Ross, no less—I was turned down. To them, the idea was too unwieldy, preposterous.

46

Motherhood changed my priorities. Before I had Nicholas, I gladly stayed at work all Saturday night when a political leadership race demanded that we close the magazine on a Sunday morning. After I became a mother, the trade-offs got tougher.

Here is what I learned: I had to create my own exits, and my own opportunities. I wanted to know my son as a toddler, and as a teenager, too. To be the mother I wanted to be, I would make compromises at work. To be the professional I wanted to be, I would make compromises at home. With those decisions came many blessings, and a couple of deep disappointments.

As a mother, I have worked full-time, part-time, and flextime: I have stayed at home and enjoyed a journalism fellowship year back at university. I did the latter when my son was two. In other words, I experimented with it all, and tried to time it well. I was entertainment editor of *Maclean's*, Canada's national newsmagazine, with a young son, and I was a vice principal of McGill University, with that same son at university himself. I could not have done the second job with a younger child.

When I look back, I see that I followed well what Sandberg advocates. I leaned in, hard. I did not leave before I left. And I made my partner a true partner: my husband and I separated when Nicholas was five, but we continued to share all daily duties related to our son, and all of the pleasures, too. As an independent television producer, based close to Nicholas's school, Will was able to respond to midday emergencies in a way I was not. We were no longer husband and wife, but we functioned beautifully as a family, and we still do. Early on, I decided I would rather have my family than a financial settlement. As a result, we didn't let lawyers get in the middle of our arrangement. We started having family dinners once a week, from the very beginning; that grew to taking shared trips with our son, and sharing cottage time. It was a novel arrangement when it began, less so now. In the end, we shared everything. The more we shared, the more smoothly things went.

Over the years, I won five National Magazine Awards for my work, and multiple others. I didn't travel a lot, but when there was a speaking engagement, I was free to go. Still, there were many opportunities I chose to forgo. More than once, the London bureau of the magazine was up for grabs. I could never apply, much as I wanted to. And I knew that without a more diverse résumé, I was unlikely to be selected as editor of the magazine, a job I once dearly wanted.

In 2001, when I threw my hat in the ring, I was not chosen. I remember being encouraged to interview for the position. My immediate response: "They aren't going to choose a woman." To which came the less-than-resounding "You don't know that." But I did. I knew it in my bones. Actually, what I knew was they weren't going to choose me.

After several rounds of rigorous interviews, the publisher poked his head in my office one noon hour. "We won't be pursuing your candidacy any further," he said, standing in the doorway, an awkward look on his face. I was eating a salad at my desk: mid-forkful, I received this news. I couldn't help but wonder: Would they tell a man this way? Wouldn't they invite him in for a short talk? Who knows, but I was scalded. An insider was chosen, and on his first day he invited me into his office. He asked me what I wanted: I said a magazine column. From there I developed a writing voice, one that gave birth to this book. It's a twisty road.

So much of this comes down to pacing, balance, and juggling—and choices. I took my career seriously; I took being a mother seriously; and for more than a decade, I took being a lover seriously as well. I had a full-time job, a vibrant speaking career, a deep connection with my son, a fabulous relationship with Jake. Always, there were trade-offs. I didn't write my column as often as I should have, in those years when Nicholas was at home. Most nights I got home late: too often, I was trundling in with groceries after seven o'clock, cooking fast for a hungry boy. In a long-distance relationship, Jake had to do too much of the traveling in the winter months. Everyone compromised. And somewhere along the line, I would surprise myself by drinking too much, using alcohol as a shock absorber.

Which isn't to say that I wasn't deeply happy with all my individual roles—mother, lover, editor, writer, speaker, daughter, sister, friend, ex-wife. But I always felt like I was failing somewhere, and I probably was. I didn't see enough of Caitlin, Jake's daughter, and that became a deep regret. More often than not, I felt stretched between multiple duties.

Most working mothers do. According to Wharton School of Business economists Betsey Stevenson and Justin Wolfers, women are less happy today than their predecessors in the 1970s, both in absolute terms and relative to men. No wonder. Between 1979 and 2006, the workweek of the typical middle-income American family increased by roughly eleven hours. According to a 2011 study by the Center for Work and Family at Boston College, 65 percent of fathers believed that both parents should contribute equally as caregivers for their children—but only 30 percent of the fathers actually did so.

For me, all the juggling took its toll. Certain disappointments at work were bruising. Menopause hit: anxiety and depression reared their ugly heads. Somewhere along the line, my occasional evenings of drinking too much morphed into drinking on an almost nightly basis. When Nicholas left for university, when the marathon was over and the house was empty, I was lonely: it was then that my evening glass of wine turned into two or three, which eventually became three or four in Montreal.

On this, I am not alone.

Preeminent American alcohol researcher Sharon Wilsnack, of

the University of North Dakota, believes we are now witnessing a "global epidemic" in women's drinking. In 2011, Katherine Keyes, now an assistant professor at Columbia University, reviewed thirty-one international studies of birth-cohort and gender differences in alcohol consumption and mortality. Her conclusion? Those born after the Second World War are more likely to binge drink and develop alcohol-use disorders than their older counterparts.

Sitting in her office, her two-year-old son's face beaming by her computer, Keyes gets specific: "Those born between 1978 and 1983 are the weekend warriors, drinking to black out. In that age group, there is a reduction in male drinking, and a sharp increase for women." Meanwhile, women who are in their forties and fifties have a very high risk in terms of heavy drinking and weekly drinking. "We're not saying, 'Put down the sherry and go back to the kitchen,'" says Keyes. "But when we see these steep increases, you wonder if we are going to see a larger burden of disease for women."

In many countries, the answer is yes. Take Britain, for instance, the Lindsay Lohan of the international set.

Most important, Keyes's study points to the critical role of societal elements in creating a drinking culture. "Traditionally, individual biological factors have been the major focus when it comes to understanding alcohol risk," says Keyes. "However, this ignores the impact of policy and environment."

The environment is challenging: witness the rise in alcohol marketing, the feminization of the drinking culture. Women need a break. They feel they deserve a break. And if drinking is about escape, it is also about entitlement and empowerment. Says Keyes: "Those in high-status occupations, working in male-dominated environments, have an increased risk of alcohol use disorders." In fact, the one protective factor for women is what Keyes calls "low-status occupations." She puts on her coat, getting ready to head home for the evening.

"As gender role traditionality decreased, the gender gap in substance abuse decreased as well. And the trajectory for female alcohol abuse now outpaces that of men."

In fact, women with a university degree are almost twice as likely to drink daily as those without. "I ask myself every day if I'm an alcoholic," says one rising corporate star, a graduate of Queen's University, who wishes to go unnamed. "I'm thirty-two, and I drink every night. All my friends drink every night. We wouldn't dream of skipping a day. We haven't had our kids yet, and we all drink the same way we did in university."

Says Katherine Brown, director of policy at Britain's Institute of Alcohol Studies: "Young professional women drink a lot more than women in manual and routine jobs—what you call blue collar. Is it marketing, keeping up with the machismo, children?" Brown believes that a crucial driver is the norms of the university years. "It's an alcohol-soaked environment," she says. "At the university I went to—Exeter—Carlsberg was a sponsor of events held on campus. The focus was on getting really, really drunk and the most horrendous things used to happen. It was an alcogenic environment—sporting events, pub crawls, often carrying a bucket around for those being sick. All social events revolved around drinking, and acting the fool was celebrated. Now, it's the 'done' thing for a city woman to come home after a stressful day and open a bottle of wine. Is it the *Sex and the City* generation? Who knows. Nobody questions it."

Walk into most social gatherings and the first thing you're asked is "Red or white?" In fact, we live in a culture where knowing your wines is a mark of sophistication. And thanks to media reports of the past several years, we have happily absorbed the news that drinking has its health benefits. For many, red wine ranks up there with vitamin D, omega-3s, and dark chocolate. If one glass is good for you, a double dose can't do much harm, can it? Actually, a double dose has its drawbacks. The largest health benefit comes from one drink every two days.

Which raises a simple question: why are we aware of the dangers related to trans fats and tanning beds, and blissfully unaware of the more serious side effects associated with our favorite drug? It's a head-scratcher, to say the least.

Last year, a study in the respected journal *Addiction* challenged the broadly accepted assumption that a daily glass of red wine offers protection against heart disease. Says Jürgen Rehm, director of social and epidemiological research at Toronto's Centre for Addiction and Mental Health and coauthor of the paper: "While a cardioprotective association between alcohol use and ischemic heart disease exists, it cannot be assumed for all drinkers, even at low levels of average intake. And, the protective association varies by gender—with higher risk for morbidity and mortality in women."

Alcohol is a carcinogen, and the risks of drinking far outweigh the protective factors. For some time there has been a clear causal link between alcohol and a wide variety of cancers, including two of the most frequently diagnosed: breast and colorectal. Rehm asks a simple question: "What would the breast cancer rate be without alcohol?"

Women have many other physical vulnerabilities when it comes to drinking. "Politically, we are equal," says Dr. Joseph Lee, medical director of the renowned Hazelden Center for Youth and Families in Plymouth, Minnesota. "But hormonally, metabolically, men and women are different—and this has implications for tolerance and physical impacts over the long run."

Women's vulnerabilities start with the simple fact that, on average, they have more body fat than men. Since body fat contains little water, there is less to dilute the alcohol consumed. In addition, women have a lower level of a key metabolizing enzyme, alcohol dehydrogenase, which helps the body break down and eliminate alcohol. As a result, a larger proportion of what women drink enters the bloodstream. Furthermore, fluctuating hormone levels mean that the intoxicating effects of alcohol set in faster when estrogen levels are high.

The list goes on. Women's chemistry means they become dependent on alcohol much faster than men. Other consequences—including cognitive deficits and liver disease—all occur earlier in women, with significantly shorter exposure to alcohol. Women who consume four or more alcoholic beverages a day quadruple their risk of dying from heart disease. Heavy drinkers of both genders run the risk of a fatal hemorrhagic stroke, but the odds are five times higher for women.

Gender is a strong predictor of alcohol use. One groundbreaking project is GENACIS—Gender, Alcohol and Culture: An International Study. With forty-one participating countries, this project offers an extraordinary opportunity to improve our understanding of how gender and culture combine to affect how women and men drink. Sharon Wilsnack, who oversees the GENACIS project, is also the lead author of the world's longest-running study of women and drinking, the National Study of Life and Health Experiences of Women. Between 1981 and 2001, she and her team interviewed the same women every five years. One of their findings: the strongest predictor of late-onset drinking is childhood sexual abuse. Says Wilsnack, "It has an increasingly adverse pattern over the course of women's lives."

Depending on a woman's stage in life, there are specific considerations of which to be aware. If you're female and adolescent, this is your brain on alcohol: consume four drinks and you will leave yourself vulnerable to compromising your spatial working memory. Binge drinking in adolescence can interrupt normal brain cell growth, particularly in the frontal brain regions critical to logical thinking and reasoning. In short, it damages cognitive abilities—especially in female teens. Says researcher Lindsay Squeglia, lead author of a study in *Alcoholism: Clinical & Experimental Research*: "Throughout adolescence, the brain is becoming more efficient, pruning. In female drinkers, we found that the pre-frontal cortex was not thinning properly. This affects executive functioning."

"Are the girls trying to keep up with the boys?" asks Edith Sullivan, a researcher at the Stanford University School of Medicine. "Quantity and frequency can be a killer for novice drinkers. Adding alcohol to the mix of the developing brain will likely complicate the normal developmental trajectory. Long after a young person recovers from a hangover, risk to cognitive and brain functions endures."

Sullivan, who has done a lot of work with the brain structure of alcoholics, is certain that what is known as "telescoping" is real: "As they develop alcoholism, women seem to develop dependence sooner than men. Drink for drink, it is worse for females."

"It is *the* issue affecting girls' health—and it's going sideways, especially for those thirteen to fifteen." This is the voice of Nancy Poole, director of research and knowledge translation at the British Columbia Centre of Excellence for Women's Health. "And the saddest thing," says Poole, "is alcohol is being marketed as girls' liberation."

Perhaps we're told too many fairy tales when we're children. From the time we're very young, we're drip-fed popular culture's notions of what will bring us happiness: being thin, being beautiful, being sexy—all these, we are told, will lead to love and success and acceptance. We already know that unrealistic images of slimness have damaged a generation. Now the alcohol industry is conspiring to drip-feed us the notion that cocktails will deliver us happy endings, rescuing us from the great modern scourges of loneliness, exhaustion, and boredom.

We need a wake-up call. For now, the first job must be jump-starting a dialogue, a fact-sharing mission. Three years ago, when I won a journalism fellowship to investigate the issue of women and alcohol, I was invited to describe my project at a media night at the University of Toronto. "I can only presume that Ann will be taking a look at First Nations women," said the worldly man introducing me. His intentions were good, but his comment was off the mark.

Last June, I had what looked to be a golden opportunity to pose some questions to Gloria Steinem. The event: a fortieth birthday party for *Ms.* magazine. I waited in line for my chance, but my audience was short. "Alcohol?" Steinem looked dismissive. "Alcohol is not a women's issue."

Perhaps not in the past—but times have changed. Whether it's a matter of escape, empowerment, or entitlement, alcohol has become a women's issue. When it comes to risk, environment and policy are key drivers of our behavior. For now, our only choice is to take a hard look at both, and face the facts. The alcohol business, like the tobacco business beforehand, has taken aim at the female market, and scored. Risky drinking has become normalized, and not all young women will mature out of it. In fact, many—like myself—may mature into it.

Here are the questions we need to be asking. Has alcohol become the modern woman's steroid, enabling her to do the heavy lifting necessary in an endlessly complex world? Is it the escape valve women need, in the midst of a major social revolution still unfolding? How much of this is marketing, and how much is the need to numb?

As a culture, we're living in major denial. It's time for an adult conversation. It's time for the dialogue to begin.

4.

The Future Is Pink

THE ALCOHOL INDUSTRY TAKES AIM
AT THE FEMALE CONSUMER

Ingredients in Bitch Fuel: vodka, gin, rum,
peach schnapps, and lemon-lime soda.
—SERVED AT RULLOFF'S, ITHACA, NEW YORK

For me, drinking was always deeply sensual. From the very beginning, I loved it all. The sound of a cork sliding from the neck of a
bottle, the glug-glug-glug of the first glass being poured, the tingle on
the tongue, and the feeling of my shoulders relaxing as the universe
seemed to say: unwind.

I loved the peaty earthiness of Irish whiskey under moonlight, the
sharp nip of Pinot Grigio at day's end. I have a particularly fond memory
of sipping scotch in the Oak Bar of the Plaza hotel in New York on a
snowy evening with Jake. Another: drinking a flute of champagne in
the lobby bar of the One Aldwych hotel in London. More than once I sat
in a Boston Whaler at sunset, dressed in lumberjack plaid, toasting the
beauty of a day's end, supremely happy as the world evolved as it should.

It is no coincidence that my favorite drinking memoir is the late Caroline Knapp's *Drinking: A Love Story*—a book I read and reread as my own drinking escalated. Wrote Knapp: "For a long time, when it's working, the drink feels like a path to a kind of self-enlightenment, something that turns us into the person we wish to be, or the person we think we are. In some ways the dynamic is this simple: alcohol makes everything better until it makes everything worse."

For more than two decades, I more or less forgot that this substance—let's name it by its clinical name, ethanol—had caused me endless sorrow and heartache in my younger years. For years, I rarely abused the privilege of drinking—and yes, I saw it as a privilege. And when my drinking caught up with me, I was as surprised and sorry as anybody could possibly be. I had promised myself I would never be outfoxed by drink.

I loved Knapp's book for many reasons, but one of the most practical was the little questionnaire halfway through the book, the twenty-six questions put together by the National Council on Alcoholism and Drug Dependence to help you decide whether you're in trouble. Knapp published her own answers, which I always thought was exceptionally generous—I felt such solidarity with her when I saw her answering "yes" to such tough questions as "Have you often failed to keep the promises you have made to yourself about controlling or cutting down on your drinking?" and "Have you tried switching brands or following different plans for controlling your drinking?" Most of all, I liked that she included the following helpful tip: if you answer yes to questions one through eight, you are in the early stages of alcoholism, which typically lasts ten to fifteen years; if you answer yes to nine through twenty-one, you are in the middle stages of alcoholism, which typically lasts two to five years. I have a dog-eared, annotated copy of her book, with my own answers marked in the margins—the early stages in 2002, and the middle stages up until my sobriety date in

2008. I am eternally grateful that I didn't make it to the third and final stage, where the questions run this way: "Sometimes after periods of drinking, do you see or hear things that aren't there?" and "Do you sometimes stay drunk for several days at a time?"

I often turned to the questionnaire in Montreal, and near the end of my stay, to the underlined portion of page 33 in my copy of the "Big Book" of Alcoholics Anonymous: "To be gravely affected, one does not necessarily have to drink for a long time nor take the quantities some of us have. This is particularly true of women. Potential female alcoholics often turn into the real thing and are gone beyond recall in a few years."

New sobriety is a fingernail-on-the-blackboard experience: many things can set you off. Restaurants walled in wine, movies with up-close-and-personal drinking shots, driving by your favorite liquor store. Billboards. Magazine ads. Just about anything. Alcohol jumps out of cupboards, into your line of vision: it has no end of tricks. You reach for ice in a friend's freezer, and there it is: a Tanqueray bottle, chilling for cocktails, taunting you.

In my case, one thing bothered me more than most. It was summer. The summer of 2008, to be exact. I was fresh out of rehab. Driving the Boston Whaler across the water, loaded with our groceries and suitcases, Jake said one prophetic sentence: "I feel like we've been studying all winter for final exams, and they've finally arrived."

He was right: summer tested me in ways I had never imagined, sitting for thirty days in treatment. Days were easy—but for me, days had always been easy. It was the waning of the light, when the magic hour approached: this was when I wanted a drink. Summer meant parties; nights with friends and family; celebrations at sunset. How was I going to handle all this?

Rehab had prepared me. I was to take a scorched-earth approach to our environment: remove all alcohol, hide all drinking paraphernalia—corkscrews, pretty glasses, ice buckets. I was to make sure my environment was alcohol-free. We asked Jake's daughter to keep any wine discreet, but we stopped short of all else. It seemed extreme. I was wrong.

Less than a week into our holiday, a phone call came late in the afternoon. A friend of my son had died tragically: an overturned canoe, a drowning, a search for the body. Jake's brother had invited us to dinner, and we were on our way there when the news came. I promised to call as soon as we arrived. I placed the call in a small den, with a fridge. My son was distraught. I was distraught, too: he was thousands of miles away, and I felt helpless.

Without thinking, I reached in the fridge for a drink, and stumbled on something new—or new to me: Mike's Hard Lemonade. Perfect: a ready-made cocktail. How convenient: vodka premixed with lemonade. I'd like to say I paused, but I didn't. I downed it quickly, without remorse, like a thief. It brought instant relief. I placed the can in the wastebasket and joined the others for dinner. I didn't say a word to Jake. I was too scared.

I snuck the occasional drink that summer: maybe twelve in total, from July to Labor Day, two on my birthday while I sautéed onions. It took the anxiety of new sobriety down a notch, and added a new worry. I knew it didn't matter that the volume was small: I had blown my sobriety. I had to go back to zero, come fall.

And by fall, I did. I gave up drinking, and stuck with it. I started going to recovery meetings regularly, and I began a new hobby: I started clipping, in earnest, the news stories that were starting to appear on women and drinking. I bought a bright yellow box and tossed in anything that caught my eye. There was not much to clip, but three stories drew my attention.

First, *New York* magazine ran an excellent feature called "Gender Bender," in December 2008. The deck read: "More women are drinking, and the women who drink are drinking more, in some cases matching their male peers. This is the kind of equality nobody was fighting for." I was one month sober. For the first time I began to think: I am not alone.

Then, in July 2009, Diane Schuler made headlines when she drove her minivan the wrong way on the Taconic Parkway, northwest of New York City. Schuler's vehicle collided with an oncoming SUV and eight people were killed, including Schuler's five-year-old daughter and three nieces, all under ten. A successful account executive and married mother of two, Schuler died with undigested alcohol in her stomach; her blood alcohol was more than twice the legal limit. Police found a jumbo bottle of Absolut vodka in the crushed metal wreck of her vehicle.

Schuler's story was horrific. I clipped it, but it wasn't the one that hit home the hardest: I never drove when I was drinking.

No, the story that got under my rib cage was one that ran less than a month later, in the Sunday Styles section of the *New York Times*, under the headline "A Heroine of Cocktail Moms Sobers Up." Stefanie Wilder-Taylor, author of *Naptime Is the New Happy Hour* and *Sippy Cups Are Not for Chardonnay*, had quit drinking. The California mother of three—best known for her popular online column "Make Mine a Double: Tales of Twins and Tequila"—had retired her corkscrew. Wilder-Taylor had announced her news on her popular mommy blog, Babyonbored, with this simple statement: "I drink too much. It became a nightly compulsion and I'm outing myself to you. . . . I quit on Friday."

What got me was this line: "Whenever her husband questioned her nightly routine, she would retort, 'I'm fine.'"

My words exactly. Whenever I had too much to drink, this was my mantra: "I'm fine." Teetering across pink granite in lake country, late at night: "I'm fine." Tossing off my high heels, after a gala awards night: "I'm fine." It was easy to say without slurring, and it was defiant. It never changed.

Except I wasn't fine. Not even close. And it was beginning to look like I was not alone. As Wilder-Taylor said, when I finally interviewed her: "Alcohol is glamorized in our society, and it's everywhere. You'd be surprised how many people are drinking during the day. And then we're shocked when some mother crashes her car with her kids in it?"

If my drinking makes me a statistic in a growing trend, it also makes me unremarkable.

I drank, for the most part, what others drank. In fact, I have no trouble charting the decades between university and 2008, using not my jobs as markers, but the wines we consumed.

In the beginning, there were those thin Italian Chiantis—cheap, astringent, and gravelly (think unreliable candleholders in walk-up apartments); plus the ubiquitous Germans (think Blue Nun). These were the wines of our university years, when we ate cheese fondue without guilt.

Of course, there was a brief stint when I—freshly graduated from university—perched for a few months in a flat in Notting Hill with the man who would become my husband. He was at film school; I was trying to become a writer. Our landlord, who played polo on Sundays with Prince Charles, had swish parties to which we were invited. There were fancy film parties, too. I remember these days with a phrase: "Mink coat, no underwear." Champagne cocktails with Helen Mirren on a Saturday night, followed by dates at Hamburger Heaven, drinking plonk. Upstairs, Veuve Clicquot; downstairs, Szekszardi.

Back home in Canada, newly employed as interns, we moved to Burgundies: easy-to-parse reds that matured quickly, just like we did.

What came next? The 1980s: travel in Europe with my filmmaker husband: fruity rosés at lunch at the Cannes Film Festival; Saint-Émilion at dinner; learning about port and the cheese tray. On my return, the Californians had arrived: Cabernet Sauvignon, Pinot Noir.

The 1990s opened with my separation and a new love affair, toasted with Australian reds, full of gusto, and golden whites. A new century brought new affluence for so many of us, and endless choice. And then? Back to the Italians: my nightly glasses of Pinot Grigio. Full circle, or almost. Where I ended was very different from where I began.

By the time this decade opened, I had stopped drinking—but I hadn't stopped watching what was going on. In fact, my bright yellow box was filling up, not just with news stories but with glossy inserts as well. What came next were the girly names: French Rabbit, Girls' Night Out, Stepping up to the Plate (label sporting a stiletto heel), and yes, MommyJuice. Wines in pretty little purse-size Tetra Paks, in picnic-perfect six-packs, ready to go.

I have at my desk three different promotional inserts from 2011. Item number one features a come-hither blonde in a sexy gold dress, balancing a martini between polished red nails, painted just a shade darker than the swizzle stick poking through the *o* in "Classic Cocktails" above her head. Call her Miss February. She's a Betty Draper lookalike posed on the front of a shiny celebration of the sixties. "You're swingin', baby!" it reads. "Do it up right like they did when after-work martinis were de rigueur . . ."

For several weeks, Ms. February was the hottest girl in town, her image stuffed into every newspaper, towering tall from storefronts. By March she was toast, supplanted by a lanky brunette in a fuchsia minidress. By April? The cover girl was no girl at all. Instead? An egg: peach-toned, hand-painted, inscribed with the name "Lily." Martha Stewart picked up where *Mad Men* left off, and a bottle of Girls' Night Out had replaced the martini.

When did the alcohol market become so pink, so female-focused, so squishy and sweet? I wondered. When did booze bags turn pastel? When did my gender become such a focus of the alcohol industry?

In 2011, I clipped a story from the *New York Times*: Clos LaChance, makers of a wine called MommyJuice, tried to get a California court to declare that they were not infringing on the trademark of a rival wine called "Mommy's Time Out." Clos LaChance argued that the word *mommy* was generic, one that no company could monopolize. Eventually, the two companies settled out of court, agreeing that both could use the "mommy" moniker.

As you might expect, Mommy's Time Out features a chair facing the corner, with a wineglass and a bottle on a table nearby. The MommyJuice label features a supple woman juggling a computer, a teddy bear, a saucepan, and a house. "Moms everywhere deserve a break," coos the back label. "So tuck your kids into bed and have a glass of MommyJuice—because you deserve it."

I called Cheryl Murphy Durzy, so-called Mom in Charge and founder of the label, at her home in San Martin, California. Why MommyJuice? "My kids call my wine 'Mommy's juice.' Lots of kids I know do this. Moms love talking about why they need MommyJuice, things like their kids wetting the bed. 'Can't wait for MommyJuice!'"

What are her thoughts about playdates with wine, about the fact that risky drinking is on the rise for women? Says Murphy Durzy: "For years, men have been relaxing at the end of the day. Does anyone ever say anything about a dad who has a beer at the ball game? No. I think it's sexist."

In Canada, the makers of Girls' Night Out wines—featuring what they call "aspirational" cocktail dresses on their labels—went to the trouble of registering their hot title in the United States, Australia, and New Zealand. Doug Beatty, vice president of marketing for Colio Estate Wines and originator of the Girls' Night Out name, says: "Eighty-five percent of the purchase decisions in the twelve- to fifteen-dollar range are 'female-driven.'" For that reason, he was "just shocked" when he learned that the name "Girls' Night Out" was up for grabs. Having expanded into wine-flavored beverages—Strawberry Samba and Tropical Tango being two—he says the future of his successful label looks "terrifyingly fun." Says Beatty, "Those of the female gender are those who have done all the hard work."

And what about Skinnygirl Cocktail line products, reported to be the fastest-growing spirit brand of 2012? Founded in 2009 by reality star Bethenny Frankel (*Real Housewives of New York*), the Skinnygirl line was the fastest-growing spirits line in the United States two years ago. Last year, Skinnygirl Cocktails—"the brand that has re-energized the way women cocktail and define themselves"—launched an advertising campaign called "Drink Like a Lady," including its first-ever national television commercial: "The lady knows how to cocktail! Skinnygirl now has all the wine, vodka, and ready-to-serve cocktails you need—without the calories you don't! Drink like a lady!" The campaign coincided with the brand's expanded product offerings, including Skinnygirl Vodka with Natural Flavors (White Cranberry Cosmo being an example) and Skinnygirl The Wine Collection.

Imbibing, without the extra calories: this is key. Last year, even the musician Fergie of the Black Eyed Peas got into the act, launching Voli, a new low-calorie vodka. It comes in six flavors, including "Original Lyte," raspberry cocoa, and pear vanilla. She was reported to have said, "I think a lot of people with healthy lifestyles like me, who love to work out, work hard, socialize, and have a drink at the end of the day, have been craving something like this," adding: "There are no extra sugars." In other words, girlie spirits, with up to 40 percent fewer calories than leading brands. Meanwhile, beer companies have started pushing their product as diet-friendly: lime-infused beverages, low-carb alternatives, lower-calorie options.

What's surprising about all this? We are used to a drinking culture pitched at men—a great example being German liquor company G-Spirits, which promises that every single drop of its alcoholic beverages has been poured on the naked breasts of a female model. Its whiskey, for instance, has dampened the breasts of Alexa Varga, Hungary's 2012 *Playboy* Playmate of the Year. All bottles come with nude photos of the model involved in the process. This example is bizarre, yet somehow predictable. So too is an ad for Belvedere Vodka, in which a beautiful model uses the reflection from a guy's shiny belt buckle as a mirror in which to apply her lipstick: her mouth is close to his crotch.

But Skinnygirl Vodka? MommyJuice? When did the female drinker become the focus of the spirits market?

I flew to Baltimore to find out. I knew that David Jernigan, the savvy, boyish-looking director of CAMY, would be willing to ball-park a date. Based in his spacious office at Johns Hopkins University, Jernigan has spent his career watching the industry. When, I want to know, did the world begin to change?

Apparently, in the late 1960s: Philip Morris, the tobacco giant, bought Miller Beer, and brought the techniques of market segmentation and lifestyle advertising to the marketing of Miller. They took a relatively regional beer and turned it into the number two brand in the United States—and they did this by using the tobacco marketing playbook. "In response," says Jernigan, "August Busch III, who was head of Anheuser-Busch, took the thick book of sporting events in the U.S. and threw it at his marketing people, saying, 'Buy this!' And they did, everything from tiddlywinks to baseball." At one point, they were sponsoring twenty-three out of twenty-four major-league baseball teams. It was a sea change: they bought into all the lifestyle marketing that had been pioneered by tobacco. Says Jernigan, "The wine and spirits folks were left in the dust."

Beer ruled North America in the 1980s and early '90s. Beer was fun, beer was sport. The spirits industry was seen as stodgy and boring. Suddenly, says Jernigan, it decided to play catch-up: it did market segmentation, looked at who was underperforming, and of course, it saw women. "For the spirits industry, this was a global opportunity. This was conscious: they understood they had to shoot younger and they had to shoot harder."

Thus was born the alcopop. Also known as the cooler, "chick beer," or "starter drinks"—sweet, brightly colored vodka- or rum-flavored concoctions in ready-to-drink format. Jernigan calls them "the anti-beer," "drinks of initiation"—and my favorite: "cocktails with training wheels." "They're the transitional drinks," he says, "particularly for young women, pulling them away from beer and towards distilled spirits. Getting brand loyalty to the spirits brand names in adolescence, so that you get that annuity for a lifetime. An obvious product for reaching this wonderful and not yet sufficiently tapped market of young women."

According to 2010 data, 68 percent of eighth-grade drinkers reported having had an alcopop in the past month, 67 percent of tenth-grade drinkers, and 58 percent of twelfth-grade drinkers. But in the 19–28 category, fewer than half had had an alcopop in the past month. Broken down by gender, the data showed alcopops were more popular with girls and women in every age group. The height of the craze for alcopops was 2004. By then they had done what the industry needed them to do—reach out to females, and establish a bridge to the parent brands like Smirnoff vodka and Bacardi rum. And of course, none of the marketing shows the consequences of drinking.

Let's take a second and look at the Smirnoff brand. In 1997, two major alcohol firms merged to form Diageo, the largest distilled spirits producer in the world—both then and now. This British-based multinational developed a sophisticated strategy to reenergize Smirnoff vodka: in 1999 it launched Smirnoff Ice, which became the number one beverage in the alcopop category. With a hefty marketing push, Smirnoff vodka sales grew 61 percent between 2000 and 2008—a sharp contrast to the 1990s, when this brand saw a dip in sales.

"Smirnoff is the girls' vodka," says Kate Simmie. At twenty-nine, she has long matured out of her Smirnoff phase: her new love is Blueberi Stolichnaya. But the McGill grad, now a Toronto marketing professional, has a firm handle on her own limits. "I'm five foot two," she says with a grin. While she has many friends who do shots, she thinks twice before joining them, or having a martini. "I can't imagine dating without drinking, but I tend to stick to wine," she says. "I can't handle shots."

Shots make a difference. Compared with distilled spirits, it takes a lot more beer or wine to produce alcohol poisoning or impairment, to compromise judgment around risky sex, which is why distilled spirits, in most cultures, are treated differently. And there's an additional health issue for women. Not only are young women experimenting with the strongest beverage, but they're more vulnerable because of the way alcohol metabolizes in female bodies. "If you're female and you're drinking spirits, and the guy's drinking beer, you're at a complete disadvantage," says Jernigan. "He's drinking a weaker beverage, he's metabolizing it more efficiently, and you're trying to keep up. And you've got Carrie Bradshaw saying that this is the image of the powerful woman—a woman with a cocktail in her hand virtually every moment that you see her, except when she's trying on shoes!"

Can we really blame Carrie Bradshaw for the martini-shots-vodka culture? Can it all be laid at her Jimmy Choos? "Let's put it this way," says Jernigan. "We cannot discount Carrie Bradshaw. But if Carrie Bradshaw hadn't been accompanied by a push by the spirits industry, she would have been a pebble in the pond. As it was, she was a boulder. Women had never been targeted before in the way they were targeted: after alcopops came distilled spirits line extensions—flavored vodkas, absolutely every fruit you could imagine."

In recent years, several countries, including Germany, France, Switzerland, and Australia, have imposed special taxes on alcopops, addressing widespread concerns about their popularity as a drink of initiation. Germany nearly doubled the tax; Australia boosted it by 70 percent. Many countries found substantial reductions in the consumption of these beverages.

And many other countries haven't done a thing. "In the past twenty-five years, there has been tremendous pressure on females to keep up with the guys," says Jernigan. "Now the industry's right there to help them. They've got their very own beverages, tailored to women. They've got their own individualized, feminized drinking culture. I'm not sure that this was what Gloria Steinem had in mind."

In the past decade, there has been a huge amount of effort to stop underage drinking in the United States. Says Jernigan: "It's made some impact with the boys. We are not getting anywhere with the girls." The more marketing kids see, the more likely they are to initiate drinking at an early age. This is 360-degree marketing, embedded in Facebook, on Twitter, on YouTube, on television, and in the movies. Last year, the Australian Medical Association censured Facebook for allowing alcohol companies to target children: "Social networking sites . . . are honing a more aggressive and insidious form of marketing that tracks online and profiles, and tailors specific marketing accordingly."

More than three-quarters of twelve- to seventeen-year-olds in the United States own cell phones; and of Facebook's one billion users, 600 million visit the social media site primarily through mobile devices. "This is the ultimate extension of lifestyle advertising," says Jernigan. "The brand is now a human being. It's interacting with you in real time. It's talking to you on Facebook. These are worlds that are being created by the brand in conjunction with, in cooperation and collaboration with, their user base. It is a marvelous innovation in marketing, and it's a disaster for us."

Brands mounting their ads on YouTube, launching their own channels: this is known as pull marketing. The consumer is seeking out the ad, rather than tuning out a commercial. They're focused. The granddaddy of this genre, Tea Partay by Smirnoff—a two-and-a-half-minute ad—has had more than six million YouTube viewers. "This is all about engagement," says Jernigan. "It's the future of marketing, and it's virtually unregulated."

There's a strong public health interest in delaying the onset of drinking: the brain is still in its plasticity state during adolescence. Every day in the United States, 4,750 kids under the age of sixteen start their drinking careers. As Jernigan says, "This is a human capital development issue."

When it comes to deconstructing advertising and the role it plays in our lives, few do a better job than Jernigan and Jean Kilbourne, the woman behind the film *Killing Us Softly: Advertising's Image of Women*. I became intrigued with Kilbourne, reading her brilliant book *Can't Buy My Love: How Advertising Changes the Way We Think and Feel*. In the opening pages, she tells the story of her own early experimenting with alcohol, which taught her that "alcohol could erase pain. From then on, for almost 20 years, my most important relationship was with alcohol." She saw a doctor about her drinking. His response: she was too young, too well educated, and too good-looking to be an alcoholic. Eventually, Kilbourne says drinking ended up "burying me alive": "I used to joke that Jack Daniel's was my most constant lover." She writes of her perfect verbal score on the SAT, dating Ringo Starr, being in love with Polish writer Jerzy Kosinski, partying at Roman Polanski's apartment—and confronting her eventual addiction in 1976. Like me, she was in the middle stages of the disease.

Astutely, Kilbourne warns us: "Advertising encourages us not only to objectify each other but also to feel that our most significant relationships are with the products that we buy. It turns lovers into things and things into lovers."

Two years ago, I wanted to meet Kilbourne, having had a spirited and bracing conversation with her on the phone. I was intrigued. Learning that she was flying to Toronto to give a speech in a nearby city, I offered to pick her up and ferry her to her destination. It was a smart idea. Kilbourne is incisive, savvy, and thoughtful. We had a long drive—good for getting to know someone, poor for taking notes: my hands were on the wheel.

The next time I spoke to her, Kilbourne was laid up at home outside of Boston, having broken her leg skydiving. I wanted to know: why are we so oblivious to the effect advertising has on us? "Ads are so trivial and silly that people feel above them," says Kilbourne. "And for that reason, they don't pay *conscious* attention. The advertisers love it: our radar is not on. We're not on guard; it gets into our subconscious and affects us very deeply."

Kilbourne quotes the chairman of an ad agency saying, "If you want to get into people's wallets, first you have to get into their lives." And there's no doubt: the spirits industry has infiltrated the female world. Which makes me want to say: Is alcohol the new tobacco?

"It took a very long time with tobacco, for people to believe that advertising and marketing had anything to do with it," says Kilbourne. "People perceive the tobacco companies as more clearly evil than the alcohol companies. Of course they're different: any use of tobacco is harmful, and that's not true for alcohol. There's such a thing as low-risk use of alcohol—although that's not the kind of use that is of any interest to the alcohol industry. If everybody drank in a low-risk way, we'd all be better off—except, of course, the alcohol companies. They'd go under. They depend on high-risk drinkers and alcoholics, and that's what people need to understand."

It may be pushing it to say that alcohol is the new tobacco, but the alcohol industry *is* the new tobacco industry. Says Kilbourne: "They've had enormous influence on politicians, enormous influence on the media, and they've framed the story: this is about your right to drink, this is about freedom. And like the tobacco industry, the alcohol industry is in the business of recruiting new users. We need to frame this in an entirely different way: this *is* a public health issue."

5.

The Age of Vulnerability

THE CONSEQUENCES OF DRINKING YOUNG

The emptier we feel, the more likely we are to turn
to products, especially potentially addictive products,
to fill us up, to make us feel whole.
—JEAN KILBOURNE

Paris, Summer of 1969

Man lands on the moon. An event eclipsed by a boy with blond hair
and an acoustic guitar.

My last night in Paris. I want to kiss him good-bye. Maybe more.

I screw up my courage with Mary Kathleen, and two
confectionary cocktails. These are our first real drinks. She wants to
kiss her boyfriend, too. Lucky for us: they're roommates.

Together, we tiptoe down the third-floor hallway of a Parisian
hotel.

Together, we fumble in the dark. Where's the doorknob?

It takes a moment for our eyes to adjust. Two beds—but not two
figures.

In fact, there are four. Naked, and asleep. Two boy-men wrapped
in the arms of older girls. Women, really.

We know these women.

These are ones who already know how to drink.

I leave Paris the next morning, without saying good-bye. I have just turned sixteen.

I didn't really drink until my university years. In high school, I did a lot of babysitting, flourished in school, and treasured my friends. Everything changed the summer I turned sixteen: my grandmother sent me to Europe for six weeks, and the world popped open like a milkweed pod. But it was brief. It all blew away when I flew home. Besides, my mother's drinking had become disturbing. I had no appetite to join her.

So when I meet seventeen-year-old Laura in a recovery group, I ask her out for coffee. All I know is that she has recently survived a small heart attack. I want to know more. I want to know what it's like to become an alcoholic before you have completed high school. I want her whole story.

We agree to meet in a coffee shop on a crisp November afternoon. A tall girl, with glossy chestnut hair, Nivea skin, and a winsome smile, she pauses and looks away several times before she can begin to talk. She lets her tea steep, wary eyes focused on the long line of cops at the register. I begin to wonder whether she has second thoughts about sharing her story.

Who could blame her? There's so much to tell. Laura, as she wants to be known, has never had a legal drink—and she hopes she never will. "There are days when I feel like I'm ninety," says the twelfth-grade student. "My friends say they think I could drink again, but I say no. I'd end up on the top of a building and wonder how I got there."

Laura tells me she had her first drink at nine, and took to it immediately. "I thought I had arrived," she says. "I remember thinking: the partygoers will accept anyone—the only requirement is to get 'lost.' I felt like I could lift buildings. And I thought: 'I'll never love anything as much as I love alcohol.'"

That love affair lasted several years. Sexually abused as a child, "debilitatingly anxious and bulimic," Laura shuffled homes, from her mother's to her father's to various aunts' and uncles'. Alcohol was a daily constant, often pilfered from relatives' liquor cabinets. "As long as I didn't have to be me, I would drink it. I felt like my skin was two sizes too small."

When she was fourteen, her stepbrother got married. She celebrated by downing nine shots of tequila. "I ruined his wedding," she says, matter-of-factly. "I threw up my body weight. My grandmother said: 'I feel sorry for you.' I didn't hear her correctly. I thought she said, 'I'm jealous of you.' That's how much I loved drinking."

The summer between grades nine and ten, she decided to turn her life around. She stopped drinking and joined a rugby team. But when she was injured, she took muscle relaxants, which she downed with alcohol. She overdosed. "My hands were yellow," she says. "My liver was failing."

By grade ten, she was taking vodka to school in a chocolate-milk container, drinking in class. At sixteen, she started stealing the anti-anxiety drug Ativan from her uncle, and buying it on the street. She loved OxyContin, did cocaine. At night she kept alcohol in a Gatorade bottle by her bed.

Her voice is very flat, but her gaze is direct. "On December twenty-sixth, I was raped by a family member. In January, I began having panic attacks, so I started mixing Ativan and alcohol. But I ran out and went into withdrawal. Two days later I was called into the principal's office. While I was there, my arm went numb and my head went backwards. I couldn't inhale. They called an ambulance. Turns out I had a mild heart attack."

When she was in the hospital, a crisis counselor asked Laura: "When was the last time you liked yourself?"

"I didn't have an answer," says Laura. "It changed my life. Two weeks later, I went to rehab."

Two months after our first interview, I attend her first-year medallion. It's a cold night, snowy and dark. She looks impossibly tall in high heels, standing at the front of the room, in a church basement in a rough part of town. With makeup, she's poised, radiant. There is no family, in the strict sense of the word. Still, her sobriety sisters are out in full force: all her roommates, many friends in the program—all young women with a past. "Thank you for coming," she says shyly, blushing at the end of the meeting. "It means a lot to me." And then she's gone, bending down for another in a long line of hugs.

Sitting with me several months later, she proudly announces she is forty-five pounds heavier than the day she had her heart attack, and her grades are 30 percent higher as well. Estranged from her family, she now lives with several sober friends and is making a documentary on addiction. She speaks in schools about her experiences, and has made formal apologies to her former teachers. "You know, girls are taken down a lot faster than guys," she says. "A friend told me: 'I don't remember last night.' I said, 'Don't you think that's a problem?'" Laura pauses. "They don't see the connection. I know girls who've gotten pregnant when they were drunk. But if you believe the Absolut vodka ads, you're going to sleep with some hot guy."

What haunts me about Laura's story is her vulnerability and her resilience, especially with news of the Steubenville, Ohio, rape wrapping up. In this case, a sixteen-year-old girl is said to have consumed a considerable amount of vodka, poured into a flavored crushed-ice drink. Much of the trial focused on the prosecution's stance that the victim was too drunk to consent to sex. Within weeks of the convictions in this case, fifteen-year-old Audrie Pott of Saratoga, California, hanged herself after photos of an alleged sexual assault were posted on Facebook—photos taken when she was passed out drunk at a party. Pott's death came within days of the suicide of seventeen-year-old Rehtaeh Parsons of Halifax, Nova Scotia, who took her life following months of bullying linked to an alleged sexual assault that took place at a house party when she was fifteen. In Parsons's case, her mother said that a photo of the alleged assault was circulated to other teens, prompting relentless torment by her peers and a steady decline in the young girl's mental health.

The tragic details of these three cases are echoing in my head when I meet with a young woman I will call Rebecca. Like Laura, Rebecca got clean and sober before she was out of her teens—and like Laura, she has had some disturbing sexual encounters. This thoughtful university student, now twenty-four and five years sober, braces herself several times before she delves into the details of her harrowing story. When she does, it comes out like a machine gun, staccato-style, her beautiful green eyes unblinking as she tells how her early drinking paved the way for other drugs, primarily cocaine and crack, and for what she calls a form of prostitution.

The eldest daughter of a wealthy Toronto family who moved a lot, Rebecca had been to thirteen different schools by the time she was seventeen. She was eleven when she decided to take a vodka bottle from her parents' liquor cabinet, and down twenty-two shots—twenty-two being her lucky number. All her friends had experienced blacking out, and she wanted to try it as well. "When you move a lot, you want to fit in," says Rebecca. "I woke up covered in vomit, and for the rest of my drinking life that was how I drank—alone and messy."

Rebecca was heavy as a child, and she was teased at school. "I developed an eating disorder, and the media had a big effect on me—how I was supposed to look, how long my eyelashes were supposed to be, how skinny my thighs should be. At thirteen, I was drinking

Smirnoff Ices, taking caffeine pills, and cutting myself." That same year, she smoked pot for the first time, but the experience was not good. "I moved to much harder drugs—coke and whatever else was available." She developed anorexia and bulimia. By fourteen, she had been sent to rehab in California by her parents. "I think I've seen more therapists than I've seen television shows," she says with a rueful grin.

With time, the Smirnoff Ice morphed into Jack Daniel's for breakfast. She was in rehab again at fifteen, at seventeen, eighteen, and finally at nineteen. Says Rebecca: "I had already uprooted so many times in my life. Then I made a lot of recovery connections—and lost them, too."

As the years went on, she started trading her body for drugs. "Everyone around me was really old—in their forties. I was sleeping with my dealer." What finally made her quit? "I hated being in my skin," she says. "As you go on, bad things happen, and you drink to forget the consequences. It's like being a rat on a wheel."

There are many consequences to drinking young, not the least of which is vulnerability to sexual assault. Says Jernigan: "If you drink before age fifteen, you're four times as likely to become alcohol dependent than those who waited until they were twenty-one; seven times as likely to be in a motor-vehicle crash after drinking; eight times as likely to experience physical violence after drinking; eleven times more likely to experience other unintentional injuries like drownings and falls. The bottom line? There's a strong public health interest in delaying the onset of drinking."

"Kids who start early are just different," says Richard Grucza, a renowned alcohol epidemiologist at the Washington University School of Medicine in St. Louis, Missouri. His response is emphatic: "Drinking early is a very strong risk factor for alcoholism. Hundreds of studies show this. In fact, there is a twenty-five percent increase in risk for alcohol dependence in those who drink at an early age."

Research shows there are influences that make a young person vulnerable to starting young. A recent British study reports that the odds of a teenager getting drunk repeatedly are twice as great if they have seen their parents under the influence, even a few times. In fact, according to Jernigan, the number one influence is the way parents drink. Ask Ali, twenty-two, why she started drinking, and she cites two reasons: because it looked like fun when her parents drank, and it seemed like a good way to deal with her crippling anxiety. At thirteen she started drinking vodka before class. At fourteen she switched to Canadian Club whiskey: "When you're fourteen, it's a joke to take CC to school in a water bottle."

Some parents believe that allowing their sons and daughters to drink at the dinner table will inoculate them against future risky drinking. In fact, a quarter of mothers in a recent American study believed that allowing their third-grader children an alcoholic beverage would discourage them from wanting to drink as teenagers: the taste would put them off. The more educated the mother, the more likely she will be what is known as "pro-sipping."

Others cite the Mediterranean model of allowing young people a taste as a way of modeling moderation. This, according to one Italian mother, is pure folly. "Young teenagers do not drink with their families at the table the way we did when we were growing up," says Tiziana Codenotti of Padua. "They drink premixed lemonade and alcohol, and on Friday night they binge. A big group will head to the square and no one knows how to deal with it."

In other words, parental influence is important, but so too are peers. In a global village, Italian youth are drinking the same way as American youth and Australian youth. All are influenced by the surround-sound marketing environment of Facebook, YouTube, movies, and TV.

Andrew Galloway, a prominent Toronto interventionist, says

younger girls are catching up to young men at an alarming rate—and they often drink for different reasons than boys. "Guys drink for the buzz and to be social. Girls drink because of lack of self-esteem, to cope, to feel a part of—and because of peer pressure. If I had a dollar for every time I heard 'I do the same thing as my friends,' I'd be a rich man. My answer? Get new friends."

"It's easy to capture the trends," says Elizabeth Saewyc, one of the lead researchers on a recent Canadian study exploring early alcohol use among adolescents. "The multimillion-dollar question is: can you capture the 'why'? But there are clear triggers," she says. "Ten-year-olds don't just voluntarily decide to use alcohol." Several key factors help tip the scales as to whether a person will drink at an early age. Number one: a history of sexual or physical abuse, or trauma. "If this is your history," says Saewyc, "you are far more likely to start at twelve or younger. If we could eliminate all violence—bullying, sexual and physical abuse, sexual harassment—we could prevent sixty-six percent of binge drinking in twelve- to eighteen-year-olds. Sexual abuse accounts for twenty percent of binge drinking, and sexual harassment for fifty percent. If we want to get a handle on problematic drinking in adolescence, we have to focus on violence in our society."

Other key factors related to early drinking: a mental health condition or a chronic physical issue, poverty, identifying as gay, lesbian, or bisexual. "They are more likely to be targets of violence," says Saewyc, "and more likely to have problems drinking." Another factor is a family history of attempting suicide. Females who start drinking at a younger age are more likely to report experiencing extreme despair, purging after eating, having suicidal thoughts, and having attempted suicide.

Those who are likely to wait until they are fifteen or older include those who are more connected to family, have friends with "healthy attitudes about risky behaviors," and have meaningful community engagement of one sort or another. For girls, two other elements are important: cultural connectedness and involvement in organized sports.

Not surprisingly, those who delay their drinking initiation are more likely to have postsecondary aspirations, some connection to a teacher or their school, and are less likely to have unprotected sex. Starting to drink too early matters for a multitude of reasons, not the least of which is that the still-maturing brain is particularly susceptible to heavy alcohol use.

In an ideal world, each young person would educate themselves on their own vulnerability to alcohol. I ask David Goldman, chief of the laboratory of human neurogenetics at the National Institute on Alcohol Abuse and Alcoholism, in the United States, about the role of genes in developing alcohol dependence. Says Goldman: "We can say because of the twin and adoption studies on alcoholism that there is a moderate to high heritability: we know that genes play a strong role. About half the reason a person becomes an alcoholic—half of the liability—is genetic." Still, says Goldman: "The strongest single predictor for both alcoholism and depression is having been sexually abused or traumatized in childhood. Genes mediate vulnerability, and there is a series of genes that affect anxiety and emotionality. If a young woman had this gene plus early life stress exposure, the probability of alcoholism increases twofold. But each gene is only a small part of the total risk. Sexual trauma is the strongest predictor."

And, as Saewyc points out, "Alcohol is the number one date-rape drug. Sexual and physical assault have an impact on hazardous drinking, and drinking to cope—which has a horrible spiraling feedback loop. Potentially, binge drinking exposes the individual to more violence. Those who have experienced sexual violence are more likely to binge drink, and this opens the opportunity to revictimization. Most will grow out of their binge drinking—but a subset will not survive, or they will have diminished opportunities. And others will develop alcoholism."

"This is what makes the human experience so different from that of the lab rat," says Goldman. "If a rat makes the right choice nine out of ten times, you are pleased. If a human makes the right choice nine out of ten times, they can fall down a well out of which it is impossible to climb. All that a young woman has to do is drink too much and end up in bed with the wrong person. The consequences can be so severe for minor slip-ups in judgment. We now have emergent strains of antibiotic-resistant gonorrhea. Globally, about half the assaults are committed by those who are intoxicated. And if a woman is drinking, she is likely drinking with a man—and likely to bear the brunt of his aggression."

If you head outside Minneapolis to Hazelden—one of the world's most renowned treatment centers—Brenda Servais will mince no words when asked for an assessment of what's happening. Says the counselor for sixteen- to twenty-one-year-olds, "Trauma? Not one hundred percent. But there's a lot of sexual trauma, whether they were sober or under the influence. They think if they were drunk, it doesn't really count because it was their fault. A lot of rape. Certainly, a lot of PTSD. And we can see a rise in substance use right after the event."

Counselors at Caron Treatment Centers in Pennsylvania agree: of their younger clients, nine out of ten report blackouts, and a lot of shame about how they got into trouble. They were completely vulnerable to sexual assault. Says psychologist Maggie Tipton, "If they're college age, sexual assault is a norm: the majority of our patients, in fact—whether they experienced it as abuse or not. Anxiety is huge." Adds her colleague Janice Styer: "A lot of young girls report that their use of alcohol is not to party. Life is hard, and it's a way to put life in the corner."

Not everyone who experiences trauma develops a drinking problem, and not everyone who has a drinking problem has trauma. But young people who have drinking problems generally have some distress or abuse in their lives. "How do we prevent the circumstances that lead to a young person thinking this is a good way of coping?" asks Saewyc. "We very seldom teach young people how to deal with stress. We teach them to read and write, but not how to heal from traumas that have happened to them, or to prevent what we can prevent. Or we teach them with a pill. Programs that do address stress have a very good effect: mindfulness in the classroom is one example. And if you're lucky, you have a family who helps you deal with these issues."

For those without that support, Saewyc helped develop a remarkable program called the Runaway Intervention Project (RIP). Based in St. Paul, Minnesota, RIP was designed to help sexually exploited or sexually assaulted young runaways—aged twelve to fifteen—put their lives back together. Young girls were being picked up when the police did drug raids. Often these girls were put in jail or juvenile detention. Says Saewyc, "The question was: what services were targeted to twelve-year-olds who had more partners than most have in a lifetime—and none by consent? Kids who run away get disconnected from school, and from caring adults."

The idea was to catch these girls early, before they became deeply entrenched in the street lifestyle. These young girls had severe health problems: the majority had PTSD; two-thirds had attempted suicide or considered it. Almost universally, they were involved in binge drinking, often in the context of sexual abuse. Says Saewyc, "Alcoholism, if not already happening, was highly likely. Substance abuse, definitely. And yet they've been able to turn their lives around."

Each young girl in RIP is returned to her family, or placed in foster care if there was abuse in the family home. A nurse practitioner makes regular home visits, helping her deal with health issues, acting as her advocate, offering emotional support and the information she needs. She is taught ways to cope with stress. The nurse practitioner also supports the girl's family. Some girls get hooked up with treatment. All go back to school and are connected with someone caring there. Says Saewyc: "As well, each girl is involved in a girls' empowerment group, with a bunch of girls with the same hairy history. The idea is to do normal things: artwork, your nails, go to movies together. We work from a strength perspective. Because they have been sexually exploited, they are at a huge risk for revictimization—and they don't always recognize dangerous people."

The girls are then assessed every three months. After six to twelve months, their trauma symptoms subside: less running away, less cutting, fewer suicide attempts. "It's common for runaways and street-involved kids to struggle with suicide because of the violence and trauma they experience," says Saewyc. "These girls get better school grades, have better relationships with their parents, and don't get pregnant as often. This intervention meets them where they're at. The most distressed girls, with fewest connections to family and school and the lowest self-esteem, improve the most. And in fact, the girls are so much better that they're indistinguishable from those who have never been abused. Last year, five of the first fifteen girls won scholarships to local community colleges."

We're living in the era of Jenna Marbles—real name Jenna Mourey. The reigning queen of YouTube, Marbles is the star of such videos as "Drunk Makeup Tutorial"—with more than 15 million views as I write—and "Drunk Christmas Tree Decorating." According to the *New York Times*, the former bartender and go-go dancer has more Facebook fans than Jennifer Lawrence and more Instagram friends than Oprah. Her fan base is the thirteen- through seventeen-year-old crowd, and she gets more than a million clicks a day for such memorable lines as "I'm going to get drunk and I'm going to teach you all the tips and tricks of how to put your fucking makeup on when you're hammered," and "People are going to be just like, 'She's drunk, she fucked up her makeup!'" Or this gem: "Sometimes you need to put some shit on your shit." Classic.

In one video, Marbles brandishes a red Solo cup, the icon of college drinking games, gets loaded, and then puts on fake eyelashes crooked. "You remind me of the girl in *Bridesmaids*," writes one young girl in the comments section. This draws a chorus of "lol."

When it comes to talent, Jennifer Lawrence is a far cry from Jenna Marbles. Still, on the night she won the Oscar for best actress, Hollywood's new "it girl" gave a shoot-from-the-hip performance in her strapless Dior gown. After a few gaffes at her post-win press conference, she confessed she had just downed a shot, giggling, "Jesus!" In subsequent interviews, she talked about wanting "to sit on my couch and drink and not change my pants for days at a time." Said the twenty-two-year-old star: "I see my couch. I see TV. I see a bottle of wine!" Perfectly honest. Still, it's hard to imagine Grace Kelly or Audrey Hepburn delivering these lines.

Times have changed. We live in an alcogenic culture. Which makes it tough to be Melanie. At seventeen, she's wrestling with her sobriety. A diminutive redheaded private school girl, sporting a silver bracelet from Tiffany's, she meets me at Starbucks a little earlier than we had first planned. She's on a tight schedule. She has tons of homework, a major presentation in two days, plus volleyball finals in the next twenty-four hours.

For that reason, we get straight to her story. When she was fifteen, her mother took her to Miami for the weekend, on what Melanie thought was a "mother-daughter bonding trip." Turns out, it was an intervention. At four thirty the next morning, two burly fellows woke her and whisked her away to Utah, where she ended up for the next fourteen months. What started with a three-month wilderness retreat ended in extended rehab. "It was traumatizing," she says. "I cried the entire time."

Melanie started drinking at twelve. "I just wanted to fit in," she says. From there, she moved almost immediately to marijuana and painkillers, stealing the latter from family members and friends' parents. "I took them from people who needed them," says Melanie. Her worst incident: she took two OxyContins with a Tylenol 3, downing them with some alcohol at school. She says: "I could have had a heart attack and died. Instead, I just puked in class—which was embarrassing. Usually, I puked in the bathroom." There's more. "I was hospitalized for three months with a very severe eating disorder—a combination of bulimia and anorexia. And my mother caught me selling drugs." She pauses. "And I got eighteen percent that year in math," says the straight-A student.

Today, she thinks about heading to university as a sober young woman, a member of Alcoholics Anonymous. She grins her shy grin and bites her lip. "It's hard. I feel like I never got enough experience with alcohol to know that I could never drink again," she says. "I feel, 'Why am I missing out on something?' I am so young. I feel like I might want to drink again—I'm an anxious sort of person, and it's sort of tempting." I ask her: what's the appeal? "I don't know how to explain it." She pauses. "I would just really like to *drink*."

And with that, she speaks for her gender—and a generation.

Binge

THE CAMPUS DRINKING CULTURE

How do you denormalize getting "shitfaced"?
—UNIVERSITY PRESIDENT

Let me introduce you to Maggie: a tall, extraordinarily attractive woman with a funky sense of style, a winning grin—and an equally winning way of speaking. Ask her about her drinking, and she shoots straight from the hip. "I was fourteen when I started, and I loved it. I had found something magical that could release the feelings of anxiety and stress I carried around. I liked getting to the point where I could be funny, escape myself. I knew I had fallen in love."

Now thirty-three and in recovery, Maggie is uncomfortable using her real name: she would prefer not to hurt her parents, or teachers who were good to her. Still, she minces no words about her high school habits, confessing that she took vodka to her girls' private school in Toronto. Once in a while, she'd down a screwdriver before heading off in the morning, or sneak one into her lunchbox. She skied, played tennis, acted, and was second violinist in the school orchestra. "But I knew I was an alcoholic at nineteen," says Maggie, who is editor in chief of a popular Toronto fashion blog. "I certainly drank more than any girl I knew." Her parents had been through a separation and a divorce that "lasted forever," one that started when she was eight. "There was a lot of sadness and confusion. I tried to take on the burden of the pain, trying to shield my little sister. By the time I had my first drink in high school, it was a relief."

By tenth grade, Maggie had already found diners that would serve her, located bootleg operations in doughnut shops. She had elaborate schemes of how to shoulder-tap at the beer store, ones that involved stuffing her bra, and flirting with any man who pulled up. Her friends also enjoyed drinking. But while they nursed Mike's Hard Lemonade in the park, she downed vodka and passed out on the grass. "I'm practically an expert on puking, hangovers," she says, looking absurdly elegant as she makes this pronouncement. She's sitting at a hip Toronto bistro, eating a chicken salad, drinking water. "I often drank until I puked, and then I would drink again. I learned how to live like that: throw up, brush my teeth, drink."

By the time she hit university—Concordia in Montreal—she says she was easily consuming forty drinks a week. "There were four or five nights every week when I was knocking back seven drinks or so. There was a bar that attracted students, one that was full of degenerate drunks. I was running a tab there. I was very depressed, and lonely. I was very unsure of who I was. I was nervous with daily things: social interaction, building a life in Montreal. If I was ever depressed in my life, it was then. But at the time, I just figured I had a drinking problem. I knew that my drinking didn't look like my friends'. Most days were spent smoking cigarettes, and most nights were spent ordering multiple double rum and Cokes."

Maggie went to the university health clinic, and the nurse directed her to Alcoholics Anonymous. "I went to AA a few times, but I knew it wasn't for me. I didn't hear anything that related to me: drunk driving, doctors drinking before operating. It didn't resonate."

Today, Maggie can take the long view on those years, having been sober for more than a year. Last summer, on an extended trip to Montreal, she returned to her college haunts: it was a "cathartic and emotionally draining" process. Looking back, what advice would she give her younger self? "Get more involved on campus," she says. "My schedule wasn't busy enough. Had I got more involved in clubs, dance classes, I would have been better off." At the time, she volunteered at a cancer ward. "It helped a bit, but I was with dying patients."

On and off, Maggie would experiment with sobriety. "There were some blocks in my mid-twenties, but I'd still binge drink every weekend to oblivion, puking on fire hydrants. I worked in the film industry. We worked hard, and played hard. I blended in."

The night before her wedding shower, Maggie—then twenty-seven—fell on the sidewalk outside a bar, cracking a tooth and ending up with a black eye and a scraped mouth. She tried to wiggle out of the shower, but there wasn't a chance: "My friends have been making party sandwiches all week," said her mother. The women who had nurtured her in childhood, all neatly arranged on slipcovered sofas, heard that she had caught her high heel in a grate and tripped. She confesses: "I looked like I'd Frenched a thorny rosebush."

Maggie may consider her own drinking unusual. But speak to McGill grad Kate Simmie, and she doesn't raise an eyebrow. "University is the acceleration of drinking," says Simmie, "not the initiation. In high school, there was definitely some stealing from the parents, drinking in friends' bedrooms. But university? People drink their faces off." The beer-guzzling frat-boy stereotype now has a female equivalent:

she drinks wine and spirits, and she's likely no stranger to drinking games. Says Simmie: "There were times when I went drink for drink with guys. Guys think that's cool."

Sydne Martin, now thirty and also a McGill grad, agrees: "The amount of vodka we drank was overwhelming. The shots of Jack Daniel's and Jäger bombs—you ended up partying just as hard as the boys did. And there was no education around it."

On Martin's first morning at McGill, someone said: "Here's your wristband, go get a beer!" "It was ten a.m., and the event was sponsored by a beer company," says Martin. "This was my first experience at university." She also remembers going on "booze cruises" of the city, with girls flashing their breasts at strangers. "I thought it was outrageous." But the incident that caused her the most alarm was one in her fourth year: a Muslim family was taking a tour of McGill, just beyond a group of frosh who had clearly been drinking and were stripping on campus, performing a series of simulated sexual acts, including fellatio. Says Martin, "I was genuinely ashamed."

Another McGill student, who prefers not to be named, said that it is not uncommon to meet young women who have woken up in the hospital, having had their stomachs pumped—both in high school and university. Recently, an acquaintance was sent home from a party in a cab. Too inebriated to pay, she was left in a snowbank by the driver. The next thing she remembers: waking up in the hospital, having had her stomach pumped. "Usually they're not embarrassed," says the McGill student. "In fact, it's often a point of pride."

Pre-drinking is the order of the day—also known as preloading or pregaming: drinking before heading out. When Dana Meyer, now twenty-five, went to Elon University in North Carolina, she would get a "handle of vodka for pregaming. In college, we drank a lot of hard liquor—whatever got you drunk the fastest." Pledging as a sorority sister, it was her job to pick up the older seniors who were "drunk and never remembered you had driven them."

Meyer knew girls in college who wouldn't eat so the alcohol would affect them faster—a phenomenon that eminent alcohol researcher Sharon Wilsnack calls "drinking efficiently." Ann Kerr, a Toronto eating disorder specialist, echoes this observation: "Their agenda is to get drunk fast. That's their intention. They may not drink and drive, but they 'pre-drink'—they get smashed before they go out. Usually, they don't eat ahead of time because it's a date. And girls tend to drink straight alcohol because they don't want the extra calories." What's changed in recent years? "The intensity of the experience," says Kerr.

"Most girls drink to fit in," says Martha, nineteen, who is heading off to college. "If you're the girl who doesn't drink, you're the loser. There's social pressure to play drinking games, guys drinking shots out of girls' belly buttons, girls chugging."

Chugging, and in many cases, not eating on a regular basis. One troubling phenomenon is the rise in so-called drunkorexia: a mixture of eating disorders and getting drunk, also known as "drinking without dining." In recent years, as the incidence of eating disorders has increased, so too has the correlation with binge drinking. According to Kerr, more than 40 percent of bulimics will have a history of alcohol abuse or dependence at some point in their lives. A recent study from the University of Missouri reported that 16 percent of college students who were surveyed reported restricting calories to "save them" for drinking. Of the respondents, roughly three times as many women reported engaging in the behavior as men. The risks of so-called drunkorexia include short- and long-term cognitive problems including difficulty concentrating, studying, and making decisions, as well as liver and blood pressure problems. Meanwhile, Dr. Naomi

Crafti—representing the Australian Eating Disorders Foundation—
reported that drunkorexia is now widespread in Australia.

"Each person with an eating disorder has their own rules about
how they get their calories," says Kerr. "The intention is to drink to
relax or unwind—and the rationale is: 'This is how I'll get my calories
today.' But these are empty calories. And the likelihood of blackouts
and seizures is very high in someone who's starved."

Eating disorders are one complicating factor. Mood disorders are
another. "Most women who have problems with alcohol also have a
problem with mood disorders—this is not necessarily true for men."
This is the experienced, compassionate voice of Kay Redfield Jamison,
author of the bestselling *An Unquiet Mind* and *Night Falls Fast: Understanding Suicide*. Professor of psychiatry at Johns Hopkins University School of Medicine, Redfield Jamison has agreed to meet me for
coffee in the campus's Daily Grind cafeteria. In a booth in the corner,
away from the fray, she is prepping for her day, already wearing her
white coat. "University is the age of risk for all psychiatric illnesses—
depression, bipolar. People don't know what they've got. This is the
worst thing: unlike Alzheimer's or heart disease or diabetes, these are
illnesses of youth. In a universe of young, healthy people, you can find
yourself sick, irritable, withdrawn. People get very agitated and drink
more. It's the worst thing you can do. It undermines medications, can
induce mixed states, make a person more impulsive—and put them
at a higher risk for suicide."

One in four individuals between the ages of 15 and 24 will experience a mental health problem, the most common being depression
and anxiety. According to a recent Canada-U.S. study, one in four students who showed up at campus health clinics suffered from depression, and one in ten had recently thought about suicide. At Canada's
Queen's University, students with diagnosed mental health illnesses
are the fastest-growing group with disabilities. Of those young people

suffering a mental health problem, only 10 percent will seek help. Alcohol, for many, is the most accessible drug.

On many campuses, alcohol has proven to be fatal. In February 2012, Zara Malone—a twenty-two-year-old student at Britain's Exeter University—was found dead in the flat that she shared, covered in her own vomit, two empty vodka bottles in her room. According to the coroner, the student of English and classical literature died "as a consequence of alcohol abuse." She had battled anorexia and insomnia, had been treated for anxiety, and had taken antidepressants. Her death was played differently from that of another Exeter student six years earlier: freshman Gavin Britton died in November 2006 after a night of playing drinking games, part of an initiation to the university's golf society that involved an extensive pub crawl. And both were presented quite differently than the death in January 2013 of Nicole Falkingham, the estranged wife of a millionaire property developer in Liverpool, who apparently died of alcohol poisoning and hypothermia in the back of a friend's car after a long stint in a wine bar.

Three different deaths, all with alcohol in common. Alcohol can kill. We often underplay this fact, beyond the issue of drunk driving. Speak to counselors in the youth group at Caron Treatment Centers in Pennsylvania and they will tell you that most young people coming in with multiple substance problems will be unconcerned about their drinking. "A lot of college students have normalized binge drinking, or drinking to black out," says David Rotenberg, vice president of treatment. Says Janice Styer, a clinical family specialist, "We see young adult women who don't even see alcohol as a problem—and parents who are grateful if their sons and daughters are *only* doing alcohol. Girls will bargain their recovery: they'll give up cocaine, but alcohol? 'What about my wedding? Graduation? My twenty-first birthday?' It's a societal thing."

Two young women, both recent graduates of Dalhousie University in Halifax, Nova Scotia, tell me that they are alone among their several female roommates in not having been arrested and put in the local drunk tank overnight. "I have no idea how I graduated," says one. "It was so accessible to get absolutely wasted in Halifax. Every night, there was a drink special at a different bar. And on Saturdays, you could buy fifty beers for thirty bucks at Split Crow. We called it 'church' because we went religiously, every week."

At Queen's University, it is not uncommon for incoming students to be told to write the phone number of a roommate or friend on their arm before a big night out: that way, should they pass out, someone will know whom to call to get them home. Says one former professor: "It's IV alcohol. The criterion is: drink to get blotto." Recent grad Rachel Shindman agrees: "Vodka is very popular for girls—they're playing beer pong or flip cup with vodka, rather than beer, drinking so much more alcohol than the guys. There are so many women who can't remember the night before—and they consider it a badge of honor."

Walking up the well-worn blue stairs of the Queen's University Health Centre, I arrive at the office of Dr. Mike Condra. Outside his office is a poster of a young woman in a black dress, lying facedown on a couch, legs splayed. A bottle of vodka is beside her, a glass tipped over. "Just because you helped her home doesn't mean you get to help yourself."

Condra, director of health, counseling, and disability services at Queen's, has been watching student behavior for years—and he notices several distinct shifts. Says the Irish-born Condra: "First of all, style. People pre-drink because it's cheaper to drink at home than at the bar. It leads people to have more alcohol in their homes. Secondly, there is the issue of quantity: it is more common to drink to serious intoxication. Students get plastered: this is part of the heart and soul here. It's that deep-rooted. I know because I have observed people who have drunk to the point of vomiting, and then gone back for more. Thirdly, there is the peer cultural influence to drink: it is considered unusual not to drink. That influence is very strong. We have an enormous amount of alcohol marketing in society, and alcohol is associated with a young, happy, positive lifestyle. And the intensity of university brings about a strong need for relatively easy ways to unwind.

"Delicately, I say it's a very difficult thing to do anything about. We can all mouth platitudes, but we need to do something serious with those who drink and do serious stuff. People forget: alcohol is a depressant. This is the prime age for the onset of most mental health problems. Drinking will delay the diagnosis. It delays help-seeking. Binge drinkers are seven times as likely as others to get sexually transmitted infections. I've seen a lot of young people make quite destructive choices. They get into fights, become sexual when they wouldn't otherwise. I've seen head injuries from falling down stairs, broken teeth, seriously broken arms. Is it a harmless rite of passage? No."

When I went to Queen's in the 1970s, it was considered uncool to abstain. Blackout drunk? This was uncool, too. Extreme drinking—getting "hammered," "loaded," "obliterated"—was for men, and only for a certain subpopulation at that: engineers and football players. Their escapades were legendary—or at least, the stuff of urban legend.

As frosh, engineers were expected to shimmy up the "grease pole": a goalpost stolen in 1955 from the University of Toronto, rumored to be coated in excrement, with a Queen's tam at the top. According to one grad, roadkill was often added to the mix, which led to infections. A Queen's yellow engineering jacket was synonymous with hard-partying macho living.

And then there were football players. One all-male house, populated by team members, was renowned for its annual end-of-year dinner party: each evening closed with the residents tossing their dirty dishes down the basement stairs. Rumor was they "washed" each other's hair with peanut butter, and urinated in empty beer bottles, bottles that unsuspecting housemates later consumed. This was pure *Animal House*: good for parties, risky for overnights.

One of my closest friends married one of those football players, a smart, charismatic young man. He eventually became the respected leader of one of the country's finest private schools. This too was part of the script: people matured, calmed down in their twenties, married, and had babies. Repeat cycle.

If this was the script, the reality was a bit different. Girls got pregnant when they didn't expect to; there was date rape; accidents happened; nice kids ended up in jail or in the emergency room. Once, in first year, I held a birthday party for my boyfriend. Someone brought margaritas, and the party took off. I ended up exiting my own party with the "bad boy." Four other couples broke up that night. That party wasn't my finest moment.

I remember only two other occasions when I drank that much in my four years at university: my then-boyfriend chewed me out, and I deserved it. But for the most part, my drinking was secondary to my primary interest, which was romance.

When Condra says that drinking has become more extreme, he is saying a lot. Queen's is renowned for three primary reasons: the exceptional quality of its students, its unquenchable school spirit, and its over-the-top partying.

Of course, there was a time in the late 1980s when the picturesque campus was also known as a hub of misogyny. In 1989, Queen's made national news when males in one residence hung signs from their windows announcing "No Means Harder," "No Means Down on

Your Knees, Bitch," and "No Means More Beer." Although this spoke more to sexual harassment than to a drinking culture, there was a strong alcohol component to the story. Female students responded by staging a sit-in at the principal's office, demanding action. The university initiated a review of both Orientation and Homecoming, and its board of trustees donated ten thousand dollars to the local sexual assault center.

Then, in the early 2000s, the infamous Aberdeen Street party—held in the student ghetto of Kingston, on the Saturday night of Homecoming weekend—began to draw crowds. By 2005 I was both the mother of a Queen's student and a senior journalist at *Maclean's*, Canada's national newsmagazine. My job: overseeing all coverage of postsecondary education.

September 2005

It's Saturday night of Homecoming weekend, and I am struggling to make my way up Aberdeen Street, cell phone clutched in one hand, notebook in the other. This was supposed to be the party to end all parties. It's a mob scene: porches overflowing, crowds swarming front lawns, pouring onto the streets, already crunchy with broken glass. Wall-to-wall people. It's only 10 p.m. and I can barely move.

Nicholas tells me he's in a dark house across the street. I strain to find the location, being jostled and shoved at every turn. "Just keep clear of the windows," I tell him. "They're lobbing beer bottles at anything and everything." "We have the lights off," he tells me. "I think we're going to head back to campus—this is no fun."

Every few minutes, you can hear the smash of glass. Police in riot gear are at one end of the street, preparing for the worst. I run into George Hood, vice principal, advancement: he is appalled that I am here. "This is all I need: Ann Dowsett Johnston reporting for *Maclean's*." He looks exasperated, tired, frantic. The crowd is building, seething. Vacant-eyed girls clutching bottles sway behind tall boyfriends in leather jackets. One girl is being treated for cuts to her feet. There's shoving, pushing; at one point, I am being carried by the crowd. The air is electric, giddy even. There is frenzy, fear. Anything could happen.

As it turned out, the mob overturned a car, setting it on fire. Police estimated there were between five and seven thousand people on Aberdeen Street that night. Mounted police tried to herd students onto lawns, with little to no success. An ambulance was blocked and police were pelted with beer bottles. The university staged a high-profile concert, hoping to distract students from the main event—to no avail. In the end, Queen's canceled its fall Homecoming weekend, with the principal declaring: "The Aberdeen Street party poses a very real threat to personal safety. It is absolutely essential that the reputation of Queen's be recognized as being based on quality, not on parties on Aberdeen Street."

Then, in 2010, tragedy: two Queen's students died in alcohol-related incidents, both falling to their deaths. In September, first-year engineering student Cameron Bruce, eighteen, fell out of a residence window. In December, nineteen-year-old Habib Khan crashed through a library skylight. When the coroner's report was released in May 2011, it called on the university to "address its culture of drinking."

Now, Queen's has decided to reinstate its Homecoming event. "I know if it goes well this fall [of 2013], we're good," says Shindman. "But if it doesn't? Well, the perception will be that the students' sense of entitlement is real." She remains critical of the extreme drinking culture. "I might be concerned about a friend who drinks six nights a week, and they will say: 'Everyone else does it.' I am flabbergasted when a person blacks out, and then is proud of it. They're putting themselves in danger, and it's celebrated! Culturally, we are not good at recognizing the problems with drinking. We waited until enough people died before we acted on mental health on campus. We need to not do the same thing with alcohol. We need to be proactive, rather than reactive."

Let's take a look at what's happening on campuses in the United States. The percentage of college students who binge drink—using the measure of five drinks in two hours for men, four for women—has been edging up in the past decade, nearing 45 percent. Consistently, this consumption level is higher than that of noncollege peers. Drinking is one thing. Combining alcohol with energy drinks is another. This habit—which produces the wide-awake drunk—has fueled many students' ability to drink harder and longer.

Each year, more than 3.3 million Americans between the ages of 18 and 24 drive under the influence. And in that same age group, an estimated 1,825 Americans die from alcohol-related causes. Alcohol is involved in nine out of ten campus rapes. Meanwhile, one in three college students meets the criteria for an alcohol use disorder. Hospitalizations for alcohol overdoses increased 25 percent for those aged 18 to 24 between 1999 and 2008.

"There is no doubt that more young women are binge drinking," says Andrew Galloway, Toronto interventionist and vice president of GreeneStone Yorkville treatment center. "We were all binge drinkers in university. Ninety percent of us toned it down. What happens to the ten percent who don't? Every young person who hears me speak thinks, 'This won't happen to me.' And a handful always end up in my office." As a generation, will this group slow down as they age? Says Sharon Wilsnack: "I am not sure they are going to 'mature' out of it. When I got into this field in the early eighties, there wasn't this huge epidemic of female drinking. Heavy alcohol use, like smoking, used to be such a male prerogative."

Thomas Workman, with the American Institutes for Research in Washington, D.C., has worked on college drinking issues for years. "With flavored vodka, the drinking of female college students became much deadlier," he says. "It is so easily abused. I still see quite a number of student deaths and egregious injuries and sexual assaults, ones that don't get publicized. Women in adolescence confuse notoriety with popularity. The fear of being the girl who will sit alone in her room and not be seen is pervasive. Somewhere, our culture has told girls that they need to be recognized—invisibility is the enemy. We see this scenario over and over. We have done a terrible job of helping young women situate themselves socially."

When he was working at the University of Nebraska, Workman got fraternities to recognize the downside of inviting first-year female students to their parties. "She is your greatest liability," Workman told them. "She can't drink well. She's your next trip to the emergency room."

Workman is still haunted by the case of nineteen-year-old Samantha Spady, a Nebraska student at Colorado State University who died of alcohol poisoning at a fraternity house in September 2004. Spady was an honors student and a homecoming queen—a small-town all-American girl who graduated as president of her class in high school. Early on a Sunday morning, after eleven hours of drinking and party-hopping that ended with her swilling too many shots of vanilla-flavored vodka, Spady was unable to stand by herself. According to various accounts, she had consumed up to forty drinks, including beer, the vodka, possibly some tequila shots; she may have participated in some drinking games. One thing is for certain: she was placed on a couch in a cluttered storage room to sleep it off. When she was discovered that evening, her blood alcohol level was 0.436 percent. Spady's parents started a foundation in her memory, one to educate on the dangers of alcohol poisoning. But as Workman says, "You need more than 'a hair-holding friend.' The fundamental question remains: 'Why on earth was she drinking so much?' "

So, how *do* you change a campus culture?

I asked that question of Rob Turrisi, professor of biobehavioral health and prevention research at Pennsylvania State University. Turrisi challenged me: "I am not sure that is the right question. How do you change the *culture*? You don't allow students who drink—and *no* campus is going to go there! Seriously, to change the culture? You'd have to change the culture of North America."

Turrisi looks at the problem differently. "You need to ask: what problem can I solve? I would ask: How many dollars are we spending on residence life, health clinics, cleanup after tailgate parties? How much is being spent in the community, fixing property damage, medical expenses, extra policing at events? Most campuses are putting Band-Aids on this problem, doing what the campuses around them are doing. Assessment might lead me to a different goal: perhaps fewer emergency room visits. If that's the case, it might be cheaper to do more triaging in a downtown facility where students are drinking.

"The administrations at most universities have been placed in a situation where they have to be a major part of the solution," says Turrisi. "Many don't have the ability—and I say this in the most positive way—to solve what is a major health problem. They were teachers, researchers, working in cell division or literature, really bright people. But nowhere in their professional development did someone give them the tools to handle this, and most of them are ill-equipped."

According to Turrisi, between 30 and 40 percent of students arriving on campus are already regular drinkers. Generally speaking, there are four types of drinkers: light or nondrinkers; weekend drinkers, who don't binge; weekend drinkers who get drunk; and heavy drinkers who have a specific weeknight they designate as a party night. Once they arrive at college, more often than not their drinking escalates. The light drinkers become moderate drinkers, moderate drinkers become heavier drinkers, heavy drinkers become heavier still. Says Turrisi: "On weekends, this last group is sort of *Lord of the Flies*–ish." These tend to be part of what Turrisi calls the "twenty-fifty" gang. Twenty percent of the population causes 50 percent of the problems: emergency room visits; blackouts; multiple repeated consequences of drinking.

Parents have a large influence as to whether their son or daughter will be in this "twenty-fifty" group: if they are loving but model heavy drinking once in a while and allow their sons and daughters to drink, their children are four times more likely to be in this group as those whose parents model moderate drinking and do not provide alcohol. Of course, there are parents who try the protective strategy of allowing kids to drink occasionally—often cited as the European model. Turrisi shakes his head: "If you allow your kids to drink once in a while, it's neutral at best—not protective."

Turrisi has developed a twenty-five-page handbook for effective parental interventions concerning drinking, best delivered the summer between going to high school and going to university. Says Turrisi: "This is a critical time for parents to parent. You are teaching self-regulatory skills to a person whose brain has not yet fully formed—and the logical, self-regulatory abilities are still developing." Turrisi says parental interventions are slightly more effective with females than males, and most effective if parents use the handbook: "It's a little more involved than just having a conversation."

Having that interchange could have multiple benefits, not the least of which is helping young women avoid sexual assault. For young women involved in binge drinking, the risk of sexual assault is relatively high. According to a recent study in the *Journal of Studies on*

Alcohol and Drugs, a quarter of young women in their freshman year said they had been sexually victimized. The more they drank, the higher the likelihood of sexual assault.

Ultimately, Turrisi believes that more needs to be spent on evidence-based approaches. Nothing surpasses the proper introduction of Brief Alcohol Screening and Intervention for College Students (BASICS), a prevention program for heavy-drinking individuals who have experienced alcohol-related problems. "You have to be very good at it to administer it well, and it takes eight weeks to be trained properly. It's good, and it works if done well. But it's time-consuming. This intervention will help students who get into trouble, but it won't change the culture."

To learn about BASICS, I turned to Brian Borsari, associate professor at Brown University's Center for Alcohol and Addiction Studies. He explains: "It's a targeted motivational intervention for students who are starting to experience consequences from their drinking: 'This is your behavior. Is this really your ideal self?' Unless you get a student to see their drinking behavior is in conflict with larger values, there is no room for change. These could be their values as a friend— they may insult friends when they drink; as a son or daughter; as a citizen, breaking the law; as a future employee, with damning photos on Facebook. It's a collaborative interaction in which the student is crafting their own reasons and means for changing their drinking— which is much better than 'You can't,' 'You should,' 'You must.' You reframe things as a choice."

One of the most exciting initiatives on the postsecondary scene is the National College Health Improvement Program (NCHIP). In 2011, Dartmouth College then-president Jim Yong Kim founded NCHIP and chose as its inaugural project the Learning Collaborative on High-Risk Drinking. Multidisciplinary teams from thirty-two universities and colleges across North America—including Yale, Stanford, Princeton, and Northwestern—embraced a model of rapid-cycle change to address the high-risk campus drinking culture. This initiative is unprecedented in both its scope, its dedication to evidence-based practices, and its commitment to move things fast. "This is improvement work, trying to take drinking down a couple of notches, with teams measuring as they go," says NCHIP director Lisa Johnson. "The cultural change is too large to shoulder alone." Forty percent of the universities have seen a reduction in emergency room visits by their students; half the schools are collecting data on a monthly basis; 65 percent implemented new brief motivational interviewing initiatives; 69 percent are surveying their students more frequently; 80 percent have implemented initiatives across campus. Says Johnson: "Have we licked this problem and figured it out? No way. It's only through sustained measures that we can make a difference."

At member institution Northwestern, five hundred students receive alcohol-related citations each year, for everything from underage drinking to vandalism in the dorm. According to Dr. Michael Fleming, professor of psychiatry and behavioral science at Northwestern's Feinberg School of Medicine, there is not one American campus that can be cited as taking a comprehensive approach to campus drinking. His 360-degree solution: "I'd screen everybody for high-risk drinking. I'd add late-night programming, offering alternatives to drinking. Dr. Turrisi's parental program is important because parental conversations work. And I'd create a community coalition: reduce the density of alcohol outlets, the number of bars offering drink specials."

Still, Fleming is optimistic about the direction things are taking. "People are finally acknowledging that campus drinking is a problem. That's a big step. Campuses are putting resources into it. There is a lot of evidence as to what works. Campuses are collecting data: real-time stuff. This is encouraging. But like smoking, it's going to take twenty years. This is a systems-level generational change."

PART TWO

On the Edge of the Big Lonely

7.

Searching for the Off Button

DRINKING TO FORGET, DRINKING TO NUMB

The central question isn't "What's wrong with this woman?"
It's "What happened to this woman?"

—NANCY POOLE

Why do we drink? To celebrate, yes. Relax, reward. Of course.

Ask most girls and women with a serious drinking problem, and you will get none of these answers. What you will get is this—present or past tense notwithstanding: I drink to numb. I drink to forget. I drink to not feel. I drink not to be me.

Abuse and traumatic stress play a major role in this reality. A Canadian study involving six treatment centers found that 90 percent of women interviewed reported childhood sexual abuse or adult abuse histories in relation to their problematic drinking. Meanwhile, the majority of young people—aged sixteen to twenty-four—in the Youth Addiction and Concurrent Disorder Service at Toronto's Centre for Addiction and Mental Health (CAMH) have histories of traumatic stress (90 percent of the females, 62 percent of the males), as well as sexual abuse.

Dr. Pamela Stewart, a psychiatrist at CAMH, puts it this way: "Typically, men drink to heighten positive feelings or socialize. Women are more likely than men to drink to get rid of negative feelings." She says that cumulative childhood trauma leads to depression and mood disorders, PTSD, and substance abuse. "Trauma is the elephant in the room. A person can present as anxious or depressed, but it can be unresolved trauma. Almost everyone with severe trauma will do something to regulate the symptoms, and in the short term, drinking does. And once you see substance abuse, you get a much higher prevalence of PTSD—ninety-one to ninety-four percent."

In the American Civil War, what we now call PTSD was known as "soldier's heart." In the First World War, it was known as "shell shock." Others knew it as "combat fatigue." "Historically," says Stewart, "we would look at trauma after each world war, and then it would be forgotten. But the Vietnam vets refused to let it slip away."

I first came across PTSD in a most personal way: I was diagnosed with it in rehab by an elegant Harvard-trained psychiatrist with a gentle manner and deep, knowing eyes. I remember having to ask what it stood for: post-traumatic stress disorder, he said, taking his time with me.

I remember being skeptical. In the tiny library, I found a book: *Seeking Safety: A Treatment Manual for PTSD and Substance Abuse*, by Lisa Najavits, a lecturer at Harvard Medical School and professor of psychiatry at Boston University School of Medicine. Taking the book back to my room, I had one thought: I want to meet this Lisa Najavits. Later, I would find the work of the wonderful Bessel van der Kolk and Judith Herman, author of the groundbreaking *Trauma and Recovery*. And of course, Gabor Maté: *In the Realm of Hungry Ghosts*. I wanted to meet them all: I had so many questions. I still do.

Journalism as therapy: it's how I've untangled a lot of my own story—depression, alcoholism. I figured if I could interview Najavits, I could test the good doctor's theory. But my first attempt to talk to her was a bust. Send your questions by email, she said, and I'll decide whether I want to answer them. We had one very awkward phone interview three years ago, in which I learned very little. It took me years to arrange to meet her in person, but meet her I did, in a private corner of the cafeteria of the Ontario Science Centre in Toronto.

First I watched her present on the subject. Standing onstage, delivering her talk on PTSD and substance abuse, Najavits shares a quote from William Faulkner: "The past is never dead. It isn't even past." PTSD, she tells us, has a very particular meaning: it has to relate to a physical event in which you were involved—the experience, threat, or witnessing of physical harm. This leaves out the stressors of divorce, poverty, neglect, and more. For diagnosis, you must have had more than a month of symptoms, causing functional problems. Problems present themselves in three core clusters: intrusion, meaning the re-experiencing of the event, flashbacks like "movie clips"; avoidance of memories or trauma discussion; and arousal, such as hypervigilance, startle reflex, and insomnia. Clients often dissociate, she says. I find myself nodding.

Clients describe using to "numb out, using to escape." In fact, most clients will be identified for their addiction, and not screened for their PTSD issues. For two-thirds of clients, the trauma comes first, the substance use second. However, for obvious reasons, substance users are at a high risk of retraumatization. Again, I find myself nodding.

Sadly, trauma symptoms do not disappear with abstinence. In fact, it often gets worse before it gets better. In sobriety, the person may get flooded with memories of the trauma. Much depends on the event: How intrusive was it? A rape, for instance versus witnessing a car accident? "Both PTSD and substance use require identity transformation," says Najavits. "Clients typically have worse outcomes because they had more positive views of their substance. And the loss of hope is one of the most profound aspects of both substance abuse and PTSD."

I find myself nodding.

Trauma-informed care emerged some time ago but is just beginning to receive widespread respect. One of the myths that Najavits is trying to dispel is that clients need to tell their traumatic story: many treatment providers think that clients need to focus on the past, using exposure-based therapy. Typically, this involves the client telling the trauma narrative using all their senses. The theory: if they do it over and over, their intense emotions will begin to wane. But it can be destabilizing.

Seeking Safety has been translated into seven languages, including Chinese—which speaks to the global need for trauma-informed tools. "It's a present-focused model for addressing PTSD—the lowest-cost model available, in that it can be peer-led," says Najavits. "The basic philosophy is: no matter what happens, there is always a way to cope safely, and use coping skills. There are millions of ways to cope safely, to recover."

What does Najavits make of the increased use of alcohol by women? "The story is still evolving," she says. "Many women believe they are not being supported in so many ways. There is a very clear connection between stress and addiction, and women are under immense stress. The more competent the woman, the more this can be hidden—and we know the nature of addiction is denial. The unseen problems of high-functioning professional women are serious."

I drank to relieve the symptoms of depression, and I drank to deal with anxiety. I wish with all my heart this wasn't true, but it is. I also drank to deal with PTSD. In other words, tough stuff happened, and it had lingering effects.

The year I turned sixteen was the one in which my mother's drinking turned nasty. It's no coincidence that my first depression hit hard that year: the feeling of too many cats sitting on my chest, coupled with a belt to the solar plexus. It was like this the first time, and it still is—if and when depression returns. And over the years, I've learned to accept the fact that it always does.

At sixteen, it whumped me hard, and I had nothing with which to whump it back. No pills, no therapy, no understanding of what was happening. I breathed deeply, slept poorly, ate virtually nothing. I wrote in my journal, read a lot of Leonard Cohen, and was utterly isolated and depleted. For the first time at school, I was affectless. I sat at the back of the class and perfected my pencil drawings of artichokes, layering leaf upon delicate leaf, all up and down the margins of my notebooks. At the end of the day, I would wander home, without enthusiasm.

One teacher reached out to me, my English teacher: she urged me to apply for early admission to Radcliffe. But this was the year of the shootings at Kent State University. Four students were killed by the Ohio National Guard. When I brought up the notion of an American university with my mother, she flatly refused. My father was out of the country. There was no discussion.

I hid in my room, wrote in my journal. I was confused and sad, without hope. And then, after eight months of hell, it lifted. It seemed like a miracle. Life regained its color; I regained my humor.

Back then, I was confused about what had happened. Maybe it was just a bad patch? Maybe it was our family's move, from a little northern town of 1,800 to Toronto? There was a pervasive sense of dislocation in our highly dysfunctional household. My depression was behind me, and I had no wish to look backward. In my last year of high school, I moved to a new school, my second in two years, made new friends, thrived. A year later, I headed off to university, made new friends in my first year, thrived.

And then it came back—only this time it was much fiercer. This time my depression had teeth. It dug in deeper. Suicide was on my mind.

This second bout of depression arrived in the heat of August, the summer in between my first and second year of university. It was a particularly bad time at home: there are things that happened that summer I am not willing to revisit or discuss. Let's just say this: it was dire, extreme. I remember feeling a desolation unlike any other, claustrophobic in its intensity. There was no way out—or there seemed not to be.

Late one night, alone in the house, I made a suicide attempt. It was feeble, but serious all the same. When I told my mother, she said: "Don't ever tell your father." That was all that was said.

Once fall came, I found it impossible to get out of bed for class. I was a residence assistant—called a don—on a first-year floor: these duties were primarily night-related, and I had no trouble fulfilling them. I would help first-year students overcome their homesickness, their heartache, their confusion. But once morning came, I found it virtually impossible to move. By early November, when the sky turned slate gray and my mood had not budged, I was desperate. I dragged myself to the student health clinic, but left before I saw a doctor: I knew the student working at the reception desk, and the stigma of seeking help was too great. I trudged home, prepared to endure another several months of psychic pain.

When the depression lifted five months later, I maintained the illusion that it had been a one-off event. Maybe I had a breakdown after all the stress of home? This is how I decided to see things. My third and fourth years of university were productive and depression-free. I dated, ran a campus arts festival, and graduated. One week later, I headed as far from home as I could possibly imagine, to Jasper, Alberta.

And that's where my life took an abrupt left turn. Working in the Rockies, in the computer room of Canadian National Railways, I met a tall, worldly man with a wicked sense of humor and a golden outlook on life. A Gene Wilder clone with a dash of Donald Sutherland, and a beatific smile. Bill to others, I called him Will. He smelled like April, when the pale green shoots of the trees reach upward, shimmering and electric. I fell in love, and so did he. He lived in Europe in the winter and worked the trains as a brakeman in the summer. Come fall, he was heading overseas to the London Film School.

That summer, the world seemed to crack open like a seed pod, full of possibilities. We moved in together, sharing a single bed in an impossibly small room. If I had my doubts, they were small ones. I was in love. And I was certain that depression was in my rearview mirror.

Six months later, I followed Will to London, perching with him in a borrowed room in a heavenly flat in Notting Hill. He had mono, and was struggling while he went to school. Still, we managed to get out and about: he had a cherry-red Mini and we zoomed around the city, parking sideways in the tightest spots. During the day, I wrote. Six months later we headed home to Canada, renting a cottage for the summer in the Lake of the Woods area. Come fall, we packed our few belongings in a bright yellow station wagon and headed to Toronto to seek our fortunes in the film and magazine businesses.

At twenty-three, I married Will. In doing so, I found a way to be adopted into a family as calm as mine was turbulent, a family filled with humor and predictability. His father was wise and had a twinkle, and we had long conversations; his mother had achieved as a community leader. I loved my birth family, and I adored my in-laws. If there were some concerns on our honeymoon, I chose to dismiss them. Life was rich.

In Toronto, Will and I began as interns in our chosen careers. We bought a beautiful, unreliable Karmann Ghia convertible, rented a minuscule apartment in the Annex neighborhood of Toronto, and rescued a stained couch from the street. In less than a year, we had a larger apartment; two years later, we had our first house. A year after that, we flipped that first home for a larger one, and a hundred-thousand-dollar profit. We bought a piece of property by a lake. We had each begun to flourish at work, he as a film producer, I as a member of the staff launching Canada's first weekly newsmagazine, *Maclean's*. We were on our way.

Were those happy years? Many times, they were. Camping on our new land, cooking dinner over a wood fire, we were in lockstep. As friends, Will and I lived life smoothly. We supported each other and were endlessly good to one another. But romantically? There was constant tension, and the tension escalated. He and I both knew, in our hearts, that we had each married our best friend. Before we had been married four years, we were in counseling.

We drank regularly in those years, and we drank well. I remember a five-week five-star trip to Europe, a trip to save our marriage. Tours of vineyards, rosé at lunch in Cannes, Veuve Clicquot at sunset, Saint-Émilion at dinner, Grand Marnier to close the evening. I began to wake at 4 a.m.: something was not right. This was too much for me. We returned to Canada and slowed it down. It was a conscious decision: we were dedicated to each other, and that included an open awareness of what had gone wrong in my family.

Six years into our marriage, we decided to have a baby. Life was on the upswing.

And then it wasn't.

Toronto, Winter of 1984

Five months pregnant, in the office. Six o'clock on a Tuesday.

Most of my colleagues have gone home. One last task before I leave: going over a book review with a writer. She and I are deeply engrossed in editing.

My boss appears at my door, pointing a gun.

I know this is some sort of twisted joke. It's not the first time he's pulled this sort of stunt. He has a bullwhip in his office, abortion tools on his bulletin board. He's unpredictable. A clever bully, with a big reputation.

"Don't point a gun at me, Alan."

"I'm not pointing it at you."

He aims it lower, at my stomach.

"I'm pointing it at your kid."

"We'll have the review to you in the morning."

He leaves. I drive home. Only then do I feel loathing turn to outrage. I call a friend. Her husband overhears the conversation. "A firearm? A guy pulled a firearm in the office?"

The next morning, I tell my editor I won't report to Alan until he removes the guns from his office drawer: target pistols, apparently. And the bullwhip. I don't mention the abortion tools on his bulletin board.

Alan denies the incident ever happened. He forgets there was a witness.

My editor insists on the status quo.

The gun incident is reported in a city newspaper. Half the office is on my side. The other half believes I was a poor sport.

My job is no longer a pleasure. I count the days until my baby is due.

Fall of 1984

Nicholas is five weeks old. Will is elated, and so am I. We head to London for a film market. For one glorious week, I retrace my steps in the city where I spent one memorable winter: up and down the steps of the Tate gallery, with my blond baby in tow; through the National Portrait Gallery; in and out of big black taxicabs with the stroller. At night Nicholas sleeps in a dresser drawer in our hotel room, one we have placed beside my side of the bed—one the maids make up with enthusiasm. Life feels complete.

On the final day, I collapse. Something is terribly wrong. In the cab home, I am on all fours.

Two weeks later, I have lost masses of weight. I have emergency gallbladder surgery. After that, postpartum depression and encroaching, terrible darkness. The constant feeling of many cats on my chest. The elevator is stuck on down.

Depression is coloring my entire world. I am weepy at dinnertime. Alone in the house with my new baby, I feel a pervasive darkness. Meanwhile, there has been a seismic shift in my feeling of comfort around my place of work. Professionally, I feel derailed and vulnerable.

Alone with Nicholas, in my nightgown. The mail arrives: Barbara Amiel, a former colleague and the future Mrs. Conrad Black, now a newspaper editor, is straddling a chair on the cover of a magazine, under the headline "Boss Lady." One thought overwhelms me: the job I once loved will never be the same.

Winter of 1985

I return to work. My instincts are correct: what was once fulfilling is openly difficult. Alan is still my boss. Within months, I apply for a leave. I win a fellowship, one that allows me to spend a year at university. The depression only deepens. On sunny days, it seems worse. Something is terribly wrong. I see a counselor.

Winter of 1986

For the first time, I find myself drinking much more than I had
expected. Not often: just two or three times. Those evenings
surprise me. I realize that I am drinking to escape. I find myself
slurring when I intend to be witty. This is not working.

In fact, very little is working: my husband has asked for a
separation; we have a beautiful two-year-old, a century-old home
with a view of the water, great jobs.

We decide to give it another try, but each day is difficult. I know
our marriage is on trial, and my future is at stake. My husband brings
me white wine spritzers, a peace offering. At first I say no. Too risky:
I know I want them for the wrong reason. But gradually, I say yes.
We begin to see a new marriage counselor, and I see her separately,
for depression. Will I take antidepressants? she asks. No, I say, I'll
tough it out. My mother's experience with Valium haunts me.

Winter of 1990

My marriage ends. Nicholas is only five. I am determined to keep
my little family together, despite the separation. On Sunday nights,
Will comes for dinner. Each evening, after we put Nicholas to bed,
we find ourselves having several glasses of wine, discussing how this
will unfold. We are apart, and still deeply connected. Nothing makes
sense.

One of the strongest predictors of alcohol abuse is childhood sexual
abuse—also related to PTSD. According to Stewart, how the abuse is
dealt with has a major effect on whether the individual will develop

PTSD. She says, "If someone steps in and protects you, it can change how you deal with the event."

In Karin's case, no one stepped in. A former advertising executive from the American Northeast, Karin is aware that her drinking began a couple of years after her older brother began doing his "sexual experiments" on her. A vibrant, petite woman in her early fifties, she pauses many times to collect herself as she tells her story, her big eyes filling with tears. Sitting in the generous kitchen of her country home, she is dressed in her fitness clothes, ready for a morning workout. Clearly, she has steeled herself for this interview. Still, it stops her in her tracks more than once. She begins slowly: "Let's just say there's a lot of shame to my story."

The daughter of an alcoholic mother, she was the youngest of four children, with much older brothers. "Our house was like a war zone—my mother didn't have the skills or capacity to be a real mother to me."

She started drinking at eleven, and was allowed wine with dinner every night. Each night, her parents would eat upstairs, and Karin would slip down to the basement to eat in front of the TV alone. "I would often get a second glass of wine, and I was allowed as long as I was quiet." She remembers first getting drunk when her beloved great-uncle died shoveling snow. She was thirteen. "I went straight to the basement and downed the better part of a bottle of wine," says Karin. "That's not what normal thirteen-year-olds do." A "binger" in university, she worked hard during the week, then did "silly, reckless things" on the weekend.

Married, with two children, Karin had severe postpartum depression, plus serious bouts of depression for the better part of three decades. But throughout her entire career, her work life has been stellar. Both times after she took maternity leave she came back to a better job and a bigger title. As she rose up the ladder, she began to drink nightly, staying out at least one night a week to drink with others from the office. "I would get home around two in the morning, and head off to work still fairly loaded—I wasn't alone in that. Once I got caught barfing in the plants in the morning—I said, 'Sorry, I'm pregnant.' The booze enabled me to cross all of my moral boundaries. It was like a lawless society: we had a strong sense that we could do anything we wanted to."

Karin and her husband always drank a lot together. "We used to pop a champagne cork out the sunroof on the way to our country place on Friday nights, arriving fairly loaded." Slowly, her self-loathing began to grow. "I used to think I was resilient, but my resilience came in a vodka bottle," says Karin. "At the end, I would have two tumblers before dinner, a bottle of wine with dinner, and more vodka afterwards."

Now fifty-three, she remembers training for a triathlon when she was in her forties, one she completed. "I'd get up every day and say to myself, 'Clearly if I can do this, I can't be an alcoholic.' Running provided enough endorphins to ease my depression, and enough structure and discipline to keep me drinking. One of my friends said to me, 'What are you running from?' I knew I was running from my drinking. I began to see a therapist. As I told her I had vowed never to become my mother, I was watching myself do just that."

What were the defining moments that got Karin to stop? One night her teenage daughter called home around 1 a.m., asking for a ride. "She was stuck at a party she didn't want to be at," says Karin. "Neither of us heard the phone. I thought she was in her bed because I was drunk." Soon after, her daughter got drunk at a Christmas party and was sent to the emergency room. Karin was too inebriated to drive to the hospital. A neighbor pitched in, ferrying Karin to the ER. "I didn't know how to talk to my daughter about this," says Karin. "I was so full of shame and remorse."

Finally, her son confronted her about her drinking. Arriving home with a gang of his friends, he found her slurring. "Do me a favor," he said. "Next time, try to be sober when I bring friends home." Says Karin, now six years sober, "That was it. I was done."

"For a while, I was very depressed. I thought I would take my life—I could take as much vodka as I could carry, and some pills, and crash my car." Instead she joined AA, quit her corporate job, and began a consulting company of her own. "At first there was a lot of relief. I was freed from the obsession of drinking. Then there was a lot of growth: I battled my depression, saw a therapist, and began to find my sea legs in sobriety. Today, I am grateful to *feel*."

Occasionally she will take a moment to have a look at the piece of paper she keeps in her top drawer: the one where she made herself write out one hundred times, "I will only have two glasses of wine tonight." There was a time when she thought that would keep her safe. It never did.

Most of all, she is sober enough to tackle the key issue of sexual abuse, and how it derailed her, and still derails her. "What sobriety has meant is that I am able to look at the past, confront it, and try to make my peace with it. Sobriety paves the way for healing."

Beata Klimek's history has no sexual abuse, but there is trauma just the same—and a history of using alcohol to numb. "You want to know about my drinking?" says the clear-eyed Klimek. "I lost my friends, my children, my mind. I did not want to be."

To me, this forty-six-year-old is remarkable: a woman willing to tell her story with unflinching candor, name included. A mother of two, comfortable sharing the details of her serious alcohol abuse, her recovery, and her life in the aftermath. This, I discover in my years of research, is exceptionally rare.

Born in Warsaw to an unwed female lawyer, Klimek is very clear on how her mother felt about her from birth: "I was a mistake." When she was very young, she was sent two hundred miles away to live with her grandparents. In six years, her mother came to visit three times— visits Klimek cannot remember. It was a happy time for her, living in a small village with a grandmother who loved her.

When Klimek was six, her mother came to fetch her. It was an unhappy turn of events: "My mother was a prosecutor, at home and at work. She was very critical, bitter, and judgmental. She never connected with me. There was a lot of emotional abuse."

New to Warsaw, Klimek pretended she was sick "all the time" to avoid going to school, where she was bullied and teased for being "fatherless." Her mother sent her to a sanatorium for six months. Klimek shakes her head, incredulous. "What kind of mother sends a young child away like that?"

In high school she played volleyball and was popular. But in her late teens she tried to commit suicide twice, ending up in the hospital. "It was really a cry for help—I didn't want to die." Her mother responded by taking away her phone privileges, forbidding her to see friends.

Throughout her childhood, Klimek saw her father rarely, but she loved him. He lived with his mother, with whom he had escaped the concentration camps. "He was an amazing, loving man," says Klimek. "He nursed his own mother for ten years when she had Alzheimer's."

As soon as she could, Klimek left home, heading off to university to study psychology for three years. When her studies were finished, she emigrated to Canada to marry, and she was soon pregnant. But by the time her son, Jacob, was eight months old, Klimek found herself on her own, a single mother working two jobs and taking English classes. Over five years she managed to put a life together. "I don't know how I did it," she says, "but I was the happiest I had ever been. I had my own car, my own apartment, a leather sofa, and my son. I was independent and I was at the top of the world."

Eventually she met a widower with a young daughter; he was a doctor. "I wasn't attracted at first," she says, "but he grew on me. He was very intelligent. I got pregnant with my daughter, Caroline, but I wasn't ready for it. That's when ten years of hell began."

At thirty-seven, she began to drink heavily. "It started with a glass of red wine just to relax," says Klimek. "I was a doctor's wife, and I was more lonely than I've ever been. We had club memberships, children in private school, everything I had ever dreamed of, but I was very unhappy. I felt trapped. My husband was critical and emotionally abusive like my mother—not loving. I married my mum. That's life: we look for something we recognize."

Klimek started to drink wine in the afternoon. "The wine helped me to unwind," she says. "And then it progressed—two glasses turned into three. I tried cocaine, but alcohol was the best because it killed my feelings."

Still, she was unprepared when her husband left her for his secretary. When their ten-year marriage fell apart, she fell apart as well. Now there were bills she could not pay. She went to Poland to say good-bye to her grandmother, who was ill, and her father, who had stomach cancer. Both died within six months of each other. She returned to Canada "broken." Says Klimek: "I had the shakes in the morning. Instead of coffee, I'd have a shot of vodka. I was still fooling some people, but not the family. I was a complete mess. My daughter was eleven and decided to move in with her father.

"I had a nervous breakdown. My way of dealing with it was to drink. I had alcohol hidden everywhere in the house. It was a medicine for me: to knock myself out. I wanted to disappear, to not feel, not think. I tried to drink myself to death. If I had to scream out all the things that make me sad, I would be screaming for days. So? I'd have another drink. I was looking for numbing, to knock myself out."

Ultimately, Klimek found her way to Toronto's Jean Tweed Centre, where she completed three weeks of an outpatient program. That experience convinced her to enter a treatment center, where she spent three months in an intensive program. Today she is reconciled with her children: her grown son sees her regularly and her teenage daughter lives with her. Still, her daughter gets antsy if Klimek has to go to the liquor store to buy something for an event. "This is a whole-family disease," she says. "Everyone suffers—especially if it is the mother."

Klimek has been sober for more than five years. Of the seven women she met in treatment, only she can make that claim. One who was a successful real estate agent is now a prostitute. Two are dead—including one who had a corporate job. Says Klimek: "You have to remember: this can be fatal."

8.

Self-Medication

MOOD DISORDERS AND ALCOHOL:
A SEDUCTIVE COMBINATION

The breeze at dawn has secrets to tell you,
Don't go back to sleep.
You must ask for what you really want.
Don't go back to sleep.
People are going back and forth across the doorsill
where the two worlds touch.
The door is round and open.
Don't go back to sleep.

—RUMI

Winter of 2002

I have decided to work on this riddle of depression by writing about it. I have sold my editors a story on depression and suicide. I have interviewed every expert I can find: endless questions, of which I never tire. I am learning a lot. I have found two willing families, both who have sons who suffered and took their own lives. The story is scheduled as a cover story at the magazine. But every time I go to write it, I freeze. My own depression deepens. I am stalled. First for voice, second for delivery. My drinking increases.

DRINK

Spring of 2005

My depression has become debilitating. I am having lunch with my wise friend Gillian. Mid-meal, I can feel the elevator plummet down. "What just happened?" she asks. "The elevator just hit the basement floor," I say. "It happens all the time." "You need to see a doctor." With her help, I see a psychiatrist within two days. My first experience with an antidepressant: Effexor.

The Houseboat, Summer of 2005

Four weeks into taking my first antidepressants, and I am losing weight at a remarkable rate. This pleases me. This pleases Jake, who is snapping my portrait daily. On a hot summer day, chopping down trees on the land he has purchased, we are surrounded by so-called sword-flies—nasty little demons that attack every available spot of bare skin. I can feel them through my thin blue jeans. "Baby, jump in the water—they won't get us there!" says Jake. We strip fast, and I hear: snap—Jake has caught my naked back as I prepare to dive into the cool Winnipeg River.

I am revving. I need very little sleep. My sister says: "I've never seen you like this." She looks concerned as I head out with an axe, armed to clear a path in the north woods. My psychiatrist says: "If you get any healthier, I fear for the forests of Canada."

I like this new energy. I begin to drink more: these pills make me rev. I can feel it in my brain. At night, I pour myself a second scotch when Jake is nursing his first. "Go easy, baby," says Jake. "I never want us to get to the point where you can't have a couple of drinks with me." My heart does a little lurch, but I choose to ignore it. I settle down beside him in the darkened night, and take a sip.

Christmas of 2005

I have hit my stride. I am finally depression-free. The suicide story
has been published. The annual university cover story has been
written. I have done 120 media interviews. I have been named a
new vice principal of McGill University and am planning a move to
Montreal in the new year. All things are possible. Only my doctor
disapproves: "It is too early for you to move away from me. You have
been in a very deep depression. Something is not right."

I like him a lot, but I think he is being sentimental. I ignore his
words.

All of a sudden, when I least expect it, I find myself having too
much to drink. An aid has become a habit.

Long before I decided to fill a prescription for antidepressants, I drank
to alleviate the pain of depression. And once I did fill that prescription, I drank to tamp down an incurable, endless revving, and sleeplessness that wouldn't quit. Antidepressants did this to me, at least
in the early months. Or maybe it was menopause: they coincided. I
would pace the halls at night, finally falling asleep past 3 a.m. Once
in a while, I'd be awake when the paper was delivered—a nasty way
to enter a workday.

Today, my condition can be managed with no trouble whatsoever.
It's a huge relief. But my years of self-medicating haunt me. I often
wonder: What would have happened, had I filled that first prescription back in the mid-1990s? Would I have sidestepped addiction? I
am wistful when I think about this: so much trouble could have been
averted.

In many women's lives, we miss the biggest part of the story if
we don't link drinking to the issue of self-medication. It's an all-too-

common reality in modern society: using alcohol for what ails us. As one beloved doctor said to me, when I used alcohol to tamp down what I call revving: "You were smart. Alcohol's a depressant, and you were using it to slow down."

It didn't feel smart at the time—but it did feel urgent and necessary. And before we get a diagnosis, or locate the correct medication and treatment, alcohol is an easy go-to substance to cure what ails us.

Women are 70 percent more likely to experience depression than men, and twice as likely to experience anxiety. And while women and men are equally likely to have bipolar I, a larger number of women have bipolar II—which typically translates to an increased number of depressive episodes in the female population, more hypermania in males.

In the United States, a woman is almost 50 percent more likely to walk out of a doctor's office with a prescription for a controlled drug than is a man. But what if she is reluctant to use pills, as I was? Or uses alcohol in concert with those drugs? Given the closing gender gap on risky alcohol use, the issue of self-medication needs to be raised. I was not alone in using alcohol to soothe troublesome feelings.

Substance use muddies the water: it makes diagnosis difficult. Says psychiatrist Pamela Stewart of Toronto's CAMH: "The art—and difficulty—of this field is to untangle what is caused by the substance and what by the underlying mood disorder."

The majority of female clients whom Cheryl Knepper sees have alcoholism with other complications: of a recent sample of 279, 80 percent had co-occurring disorders, and 85 percent had sedative dependence. Knepper, vice president of continuum services at Caron Treatment Centers, says many women are adding Valium, Xanax, or Ambien to their drinking. "Life starts to hit them and prescription drugs become their best friends, along with the alcohol. They don't share with their primary physician that they are using alcohol to self-soothe. Couple pills with alcohol, and it's a slippery slope. Women have a tendency to want to project an image of holding it all together—but they will know internally, long before others, that they have a problem. Eventually they will be outed: a DUI, showing up late for work too often, external issues. And by the time they come to treatment, the majority are here because their family asked them to."

Self-medication: of all the women I met while researching this book, Alex was one who used alcohol to treat full-blown bipolar I. Hers was also one of the most compelling life stories: certainly the one with the most dramatic end-of-drinking tales.

Sitting on a plush couch in her penthouse suite, Alex confesses that to this day, she is not sure what really happened on that plane to Las Vegas. Did she really take her top off in first class, as the rumor goes? Did she actually punch the male flight attendant in the groin? Who knows. All she's really sure of is that she boarded a flight in Chicago, drunk, en route to a major sales conference.

Owner of several successful businesses, and president of her industry's North American association, she was well recognized on the flight. She has a vague memory of lying on the floor at the luggage carousel. When she came to, out of her blackout, she was unpacking her clothes in her hotel suite. Minutes later, there was a knock at the door, and a colleague arrived with a large jug of Kentucky bourbon. The next morning it was breakfast.

This was Alex's last hurrah—and a very public one at that. On the conference floor, an old friend confronted her: "What did you *do* on that plane yesterday? Everyone is talking." Turns out, airport security had escorted her to the luggage carousel, where she had passed out. Those who recognized her persuaded security to release her. Says Alex: "In a blackout, I got myself from the airport to the hotel room.

By that point, I was a *functioning* blackout drinker. I ended up drinking for the next four days."

Impeccably fit, confident, and articulate, Alex sold her three companies ten years ago in a series of multimillion-dollar deals. The Las Vegas incident is now twenty-two years in the past, and she is generous with the details concerning what led up to that fateful trip.

The pivotal incident in her childhood was the death of her younger brother: hitchhiking at the age of thirteen, he was killed by a drunk driver. Alex was fifteen, and "all hell broke loose in our family." Her mother was sedated; her father's drinking escalated. "There was no grieving or consoling as we know of it today, no hugging. It was very brutal. It certainly changed things in my family." Most of all, Alex felt incredible guilt: she had noticed that night that her brother wasn't in his bed, and decided not to tell her parents. She felt "quietly responsible" for his death.

Within months, she had her first drink, and experienced her "first of hundreds of blackouts." Downing two-thirds of a bottle of rye within a half hour, she threw up everywhere and ended up with a huge bruise on her face. The next morning, uncertain of what had happened, she put on her tennis clothes and headed to school. The incident was all over the school. "It was complete humiliation," she says. "I was honor athlete of the year, Miss Goody Two-shoes, and it turns out two guys had had to hold me over the toilet for most of the night. I didn't drink again until university."

At twenty-three, she launched herself in the family business, working for a company based in Dallas, a division of a multinational. There she did a one-week boot camp in sales, and then stayed on the road for the following year. Seven days a week, she stayed in hotels and motels, some so sketchy she slept with the lights on. Within six months, she had increased sales by 400 percent. "I loved the product, I loved the work, I loved my customers. I was in heaven."

She was also drinking daily. "I would drink on planes, drink in hotels. I drank to blackout all through my twenties, and it would not be unusual for me to take someone home—and not know them in the morning. I was doing all sorts of inappropriate things with inappropriate people."

While Alex built a very successful company, it was fueled by undiagnosed bipolar disorder. "My creativity, ability to negotiate, and reserves of energy were when I was really manic. I was very speedy." She negotiated deals until 3:30 a.m., chartering planes in the middle of the night, all stocked with full bars. "I was hugely manic. I would get to that agitated state and I could do eight more things than I am usually able to do."

At thirty-four, she met Ian, a very successful older man, who said: "How can I help you make your business grow?" He helped her arrange the financing to buy two companies, and the two fell in love. Their honeymoon was spent on the road, working. Most of all, they drank together—a huge amount. "Two martinis before dinner, two bottles of wine with our meal, and then something else afterwards."

"Ian was my biggest cheerleader, and we sold our companies at a profit." Together they bought a twenty-thousand-square-foot lakeside home, and began spending every weekend there, traveling on Fridays, each with their mug in hand—his of gin, hers of vodka. "It would just be a lost weekend. We would isolate with our books, magazines, newspapers, and alcohol."

Then, on New Year's Day 1990, Ian woke and declared, "I think we should quit drinking and smoking." He did; she did not. She would sneak into bars, downing double vodkas. "I worked very hard on that marriage, to make him love me, but it ended by September." Alex's drinking escalated. "No one was watching, at that point," says Alex. "All I could think of was: 'I managed to fuck this up. I might as well drink.' I would take a vodka bottle out of the freezer and while the bottle was in my mouth, I would lecture myself. And that was the beginning of the day. I would get a block from the office and realize I was too messed up to work. I would lie about why I wasn't coming in, take down my messages, and pass out in the car for a couple of hours—at ten or eleven in the morning. Later, I wouldn't be able to read my own writing. I was incapable of showing up at my own business. I would book appointments for three in the afternoon and hope that I could make them. For months and months, I was incapable of functioning."

Not only did she drive drunk: at that point, she also had a horse and she would ride drunk. Increasingly, others became worried. "My father called my GP. My staff knew. I never went anywhere without a ton of vodka—and the only time that vodka doesn't smell is in the bottle, with the cap on!"

The week after the Las Vegas incident, Alex's senior executive asked for a meeting in the boardroom. He said: "You may fire me for saying this, but there are a lot of people who care about you who are very worried about what happened in Las Vegas."

Alex responded by driving to her parents' cottage drunk, picking up two hitchhikers on the way, stopping at an old boyfriend's cottage en route, finally arriving at 2:30 a.m. "I was driving on the same road where my brother had been killed." The next morning, she downed a mickey of vodka and visited one of her sisters, who said: "I am not going to lose another sibling to alcohol. You are going to treatment."

Alex phoned the Betty Ford Center drunk, securing an intake date three weeks hence. In the meantime, she drank every bit of alcohol she owned, with the exception of her Tia Maria. "My sister dropped me off at the airport, saying: 'I can't imagine you never having another vodka martini or glass of wine.' I said: 'Me, too.' With that, I boarded the plane and ordered a tumbler of vodka in first class."

Still, she was sober when she landed. Within days, she was introduced to the Twelve Steps. "The compulsion and obsession to drink was lifted," says Alex. On the first anniversary of her sobriety, she treated herself to a black Ferrari. But she was not contented. Quite the opposite: she was depressed and anxious. Without alcohol in her system, the symptoms she had grown used to were much more evident. "I hit bottom in sobriety. My system would flip every three to five weeks, from mania to deep depression. Other people were getting more serene, and I was not. Half the time, I wanted a twenty-two-wheeler to roll over me, and the other half, I could tell people thought I was on cocaine or speed. When I am offensively manic, I want to go shopping and be in public. All I have to do is look at my promiscuity or my jewelry"—she holds out a spectacular diamond ring—"and I know when I am manic. I get so high and manic that I feel like the ball in a pinball machine on full tilt—nobody does anything fast enough for me!"

Five years sober, she was told by her doctor that she was manic-depressive and should go on Lithium. Alex was diagnosed with rapid-cycle bipolar, discovering that much of what had fueled her remarkable productivity and her rocketlike career was her disorder. "To be diagnosed with this was as big a relief as it was to be diagnosed a plain-vanilla alcoholic," she says. "Alcohol had taken the edge off the very high points and off the very lows as well. It was clear I had been self-medicating. The symptoms were greater sober, and all magnified. The more sober I got, the more severe and violent they were."

Today, Alex is dedicated to routine. Alcoholics Anonymous is an enormous part of her life. She says: "When I am not on the beam in my program, it is as if my twelve-cylinder Ferrari is in third gear at sixty-five hundred rpms—I need to take my foot off the pedal and breathe." In addition, Alex works out three times a week, plays tennis and golf, and travels around the world. "I lead a very charmed life," she says. "I sold my last business in 2003, and I did really well. It has allowed me to be very generous with my family, and my foundations. And if I weren't sober, I wouldn't have any of this."

As I was writing this book, I knew that I wanted Marion Kane's story—hers being one of two addictions, intertwined. Still, it took her several months before she decided to share her story with me, using her real name and all her history, unvarnished and undisguised. When the well-known food writer finally welcomes me into her living room in Toronto's funky Kensington Market area, she begins to speak with a journalist's flair. "They conspired, my two addictions: for almost thirty years, they were benign and somewhat separate. Then they weren't, and they came together in one demonic combination." Her eyes are wise and frank, and she uses her hands when she talks. I like her immediately.

Kane first took Valium in 1976, when her ten-year marriage, to a man she had known since she was fourteen, came apart. "It was a nasty breakup," says Kane. "He left precipitously, and our daughter was only five. I was overwhelmed by motherhood. This breakup unleashed abandonment fears, ones I'd had since childhood. I realized that I had been suffering from depression since my early teens. But the world went really gray when I was thirty. That year, I started drinking with the pills—which multiplied the effect of the sedation. I can remember falling out of a cab—I was starting to self-medicate."

By the early 1980s, life had turned around for Kane. She started her food writing career, a vocation she loved. "I was the food editor at the *Toronto Sun*, not taking pills, happy because I had found my passion." By that time she was in a second relationship, and in 1987 she had a second daughter. A couple of months later, she and her partner broke up. "Being a mother again and having a career was the happiest time of my life," she says. "I still had episodic problems sleeping—I had anxiety as a child and a lot of insomnia—but I am a feisty individual. I had a good happy period for several years."

In 1989, she was wooed by the *Toronto Star*, Canada's largest newspaper, to become their food editor. "I believe I am an adrenaline junkie," says Kane. "I loved my job. I interviewed them all: Julia Child, Jamie Oliver before he was big, the chef who escaped from 9/11. It gave me an outlet for my creativity, and for about fifteen years, I was happy."

Still, she was taking Imovane and lorazepam to sleep, rotating them. Then menopause hit, and her sleep problems grew worse. "I realized that I might sleep better if I had a glass of wine with the pills," says Kane. Eventually she discovered that brandy worked very well, too. "I would take a few shots of brandy, a couple of pills, maybe another few shots in the middle of the night, and more pills." Kane had episodes of falling down, the odd bloodshot eye, a stomach ulcer, gallbladder issues.

The rubber hit the road when her younger daughter turned seventeen and left home. "I had trouble at work, an empty nest, menopause all at once," she says. "Menopause is an upheaval of the first order. I was an addict, waiting to happen."

Thinking she could escape all this, Kane experimented with a geographic cure: she moved to the small town of Stratford, Ontario, buying a big house on a quiet street. But despite the fact that she made several good friends, she was isolated and missed the urban scene of Toronto. "Soon I had started to drink copious amounts and take pills every night at bedtime. One night, I was on assignment in Miami, for the South Beach Wine and Food Festival. I was staying at the Fontainebleau hotel, with a balcony and a view of the ocean. I had a headache, so I took three Imovane and drank four or five little rum and vodka bottles from the minibar—I had a great sleep. I started taking the vodka bottle to my bedroom. At first it was a couple of nights a week, and then it was every night. Often I would have 'lost weekends,' sleeping the whole of Saturday and Sunday. It was incremental and gradual. My goal was oblivion."

On more than one weekend, Kane booked into one of Toronto's better hotels, hiding behind the Do Not Disturb sign. One weekend, her youngest daughter came searching for her mother and discovered she had consumed the whole minibar. Says Kane, "I was in denial. I genuinely didn't think I needed help."

In 2008, a doctor told Kane: "You will die if you continue." "That was the date of my last drink," she says, "but it took six weeks to get off the pills. The withdrawal was horrendous. Each week was harder than the last. For four months, I would go to my friend's house and say, 'Help me!' The withdrawal was made worse by an antipsychotic I was prescribed to help me sleep, called Seroquel. I was going from bad to worse. Every morning I would wake with a feeling of impending doom."

Finally, she went to rehab. "It's been a long, bumpy road of recovery," says Kane. "Rehab was like boot camp, but it saved my life. It wasn't a picnic being there. I hated it at first, but there were AA meetings and I was with kindred souls. I was with a lot of soldiers who had come back from Afghanistan and Bosnia. When I arrived, I couldn't peel a banana or make a cup of coffee—people thought I was a junkie. But they nurture you. Being parented, cared for: this is what you need.

"I want people to understand that we don't become addicts because we want to destroy ourselves," says Kane. "It's about banishing the demons—and I had a lot of demons. I have too many antennae—it's a curse and a blessing. I'd like to have a dimmer switch, to turn things down."

Kane believes that most addicts have been deprived of love as children. "My family was pretty dysfunctional," she says. "My mother, who is a biologist, fled the Holocaust. Most of her family was murdered. Both my parents had had brutal lives. I grew up in a secular Jewish household in postwar England—both parents had funny accents, we ate funny food. My father, who taught medicine at University College London, was a very angry guy, although brilliant and a softie at heart. It was a chaotic household."

Kane is radiant as she pours tea, and tells me of her life today. "I have healed a rift with my eldest daughter. I live with a good man who treats me well. And I have really made my peace with my mother, with whom I never got along. I believe that recovery is a process of being reborn. You know, you can keep stumbling, or you can go through a crisis and find deeper meaning and peace. Today, I take no drugs to sleep. I exercise and I try to find peace of mind." What's her secret to sleeping? "I watch Jon Stewart each night before bed." She tosses back her head, and laughs her throaty laugh. "A lot has come together."

Sleeplessness and alcohol use: it's a common coupling. Anxiety and alcohol is, too, and to discuss this I turn to Julie—or we'll call her that. A raven-haired woman in her early thirties, with enormous brown eyes and a delicate tattoo etched on her inner wrist. Very hip, very open, well-known on the Toronto scene. "With me, it always comes back to anxiety," she says over grilled chicken in an outdoor bistro. "I think I've had an anxiety disorder my whole life. As a child, it would manifest this way: I couldn't sleep because I had such a persistent fear that our house was going to catch fire. How would I save my little brother and sister? 'I am going to kill them somehow. I am just not going to be a good sister.' That's what used to haunt me. God knows where that pressure came from."

Julie started drinking when she was eleven: "Everyone else was drinking, so I did, too." It was during those years that her anxiety escalated and she began having panic attacks, crippling ones by the time she was sixteen and seventeen. They happened often, especially when she drove. "The more I thought about it, the worse it got. I went to a doctor, and took pills. But the only thing that seemed to work was alcohol. Alcohol was the only thing that would get me through minute by minute."

When she was in her late teens, her parents moved to another city for work opportunities, and Julie moved into her own apartment. By then her anxiety had turned into full-blown agoraphobia. She would drink just to be able to go out to a concert—and then black out by the second half.

Once she finished college, her friends started settling down—but Julie continued to party at bars. "It was like everybody else went home and I stayed out." She gets very quiet. "In my mid-twenties, I was with the man of my dreams, shopping for rings, looking for condos. He came home one day and said, 'I can't do this.' Why did he leave me? Things were getting bad. I would buy two bottles of wine—and finish one before he got home from work. After he left, my drinking escalated. The times I was feeling terrible were getting longer and longer. I saw no hope and no answer.

"I would put myself in situations that no one would dream of. Drinking a bottle of wine, and then driving to get another—I will never forgive myself for that. Nights were longer on red wine or beer. On vodka, they were shorter—vodka my blackout drink. One night I went to a vodka bar. The next morning I woke up in a hotel. There was a pair of boxer shorts on the floor. The guy was nowhere in sight. I didn't know where I was. I called my roommate and said: 'Where am I?' She said, 'You're kidding, right?' "

Today, Julie is on anti-anxiety medication, sees a doctor regularly, and just picked up her one-year medallion at AA. Without alcohol, she's thriving: a new job, a new man in her life, and a calendar filled with international travel. "I still get anxious," she admits. "But I can fly, I can see the world, I can go to a concert without freaking out. It's *so* cool." She grins her impossibly charming grin, and I have no trouble believing her as she heads off into the soft summer night, her bag slung jauntily over her shoulder and her head cocked toward the light.

9.

Romancing the Glass

A SLIM STEM OF LIQUID SWAGGER

There is some kiss we want with our whole lives.

—RUMI

Toronto, Spring of 1996

A party on a cold Wednesday night: I am reluctant to go, but my friend Victor Dwyer persuades me to get a babysitter. What is purported to be a fiftieth-birthday party for my friend, writer Marni Jackson, turns out to be her surprise wedding celebration: she has married her longtime partner, film critic Brian D. Johnson. Zal Yanovsky—formerly of the Lovin' Spoonful—is onstage, pumping up the crowd. Brian is on the bongo drums: delirious tribal joy. I feel self-conscious in my singlehood.

A handsome man approaches me. "Remember me?" "Of course. You're Jake MacDonald, from Winnipeg." Almost two decades earlier, we had had a twenty-minute chat at a party, talking about writing. I sat on a piano bench, a newlywed, happy with my new magazine job, talking to an aspiring writer. I wore white linen and brown velvet. He had beautiful eyes, dark hair. I remember him well.

"Can I get you a drink?" It is me offering: he doesn't know

this crowd. He has come with his good friend, the writer Paul Quarrington, and he's looking to me for cues. "Sure, a beer," he says. He thinks I'm finding a way to ditch him, but I return with two drinks. We begin to share life stories: our failed marriages, the fact we have children the same age, our mutual interest in writing.

After a long time, we part. But when the music starts, he crosses the floor and asks me to dance. It's a slowish song, which catches me off guard. Something electric happens when he holds me in his arms: I feel a current in my whole being.

We find a quiet corner at the bar. "I have a question," he says. "From where I sit, it looks like you have everything in life—a great job, a son you love. For a woman with so much, why are you so sad?" I tell him about lost love. He has deep, questioning eyes. He's a good listener. I am, too. I learn that the second of two marriages has just ended. I am intrigued.

When the last guest leaves, we are still perched on the bar stools, cleaners sweeping around us. Outside, in the crisp night air, standing in our trench coats, he offers to escort me home. In the middle of negotiating who will drive, we kiss. For the next fourteen years, our lives will be intertwined, transformed, blessed.

Two nights later, both of us are headed to the National Magazine Awards. I dress with extreme care: my sheerest stockings, my slimmest skirt, my favorite perfume, lipstick. When I leave the house, I tingle.

Hours later, we arrive home, two writers flushed from winning. His magazine has given him a bottle of scotch. Will I keep it for him, to share the next time he is in town? I blush. Of course. I pay the babysitter, pour him a glass of Irish whiskey, and the story begins to unspool, like a fairy tale—only better.

The Houseboat, Summer of 2003

In the lake, both naked, mid-morning. Jake and I, treading water, talking. I look beyond his head and I see two things approaching: a moose, and a Jet-Ski. The moose will retreat. The Jet-Ski won't, bearing as it does a man who has traveled for an hour to find the houseboat, to tell Jake he wants a book autographed. Jake has won a national award for his book *Houseboat Chronicles*, and his fan club is growing. It makes me deeply happy. But Jake tells the man to come back another time: he is not going to interrupt our swim or invite a stranger into the houseboat. I think: what a singular man.

That night, I am inscribing a book for him, one called *The Sea*. Carefully, I copy the words of a poem I love: "Natural History," by E. B. White, one he wrote for his beloved Katharine in Toronto's King Edward Hotel. A poem that says too much about the fact that Jake and I live in two different cities and can't resolve where to call home. We each have a child, there are two other parents involved, and three places we call home. This is complex. Both of us are writers, and the web we are devising will be our eventual undoing.

But for the moment, I am determined to see the complexity of our arrangement as a gift, a perpetual romance. And so I transcribe the words of White, a talisman to ward off the curse of long-distance love:

> The spider, dropping down from twig,
> Unfolds a plan of her devising,
> A thin premeditated rig
> To use in rising.
>
> And all that journey down through space,
> In cool descent and loyal-hearted,

She spins a ladder to the place
From where she started.

Thus I, gone forth as spiders do
In spider's web a truth discerning,
Attach one silken thread to you
For my returning.

Summer of 2005

The days are perpetually golden. Fresh-picked blueberries for breakfast. Frequent swims. Long, meandering conversations over meals at our little driftwood table, dinners of fresh pickerel—fish we caught ourselves. Each evening, as I chop the vegetables and sip my wine, Jake creates a playlist, calling out in his deep, gravelly voice: "Can you hear it, baby, or do we need it louder?"

Before bed, sipping scotch under the stars, we watch the northern lights, listening to our favorites: Van Morrison, Mark Knopfler, Keith Jarrett. In the morning, we wake to the sound of otters raiding the minnow bucket. "Those little bandits!" says Jake, leaning over me to check out what's going on. His arms are the color of mahogany. He lies back down and says what he always does: "Where were you all night? I kept trying to find you, and you kept trying to escape." He rolls over to kiss me properly. Another day begins, heavenly.

This is the last summer that the little houseboat will sit on the water: Jake has bought a piece of land, and the houseboat needs to be moved up the river to its new home. We wait for a calm evening to make the voyage, multiple boats towing our romantic hideaway to its new destination. Finally, the right evening comes. Slowly, we make the voyage upriver, pushing off from Virgin Island, around the corner into open water, and under the railway bridge. We toast this auspicious occasion by opening a bottle of white wine, then another and another: There is a large crew. Some are drinking gin and tonic. I am sitting with Jake's daughter Caitlin and her friends, pouring wine, telling stories. There is a sense of foreboding that I cannot shake. That night, Jake and I hold each other very close, as we always do, spooned together, tight. Much remains unsaid.

Within a week, we will be in Paris, where I am giving a keynote address at a major conference. Night after night, we sit in little French bistros, designing our new home on paper napkins—the houseboat as it transitions to a "landboat." We decide to put a fireplace in our bedroom, to fashion a room just like the one we love at Kamalame Cay in the Bahamas. While I deliver my speech, Jake sources a favorite lamp to give me as a gift. We reunite in our tiny bedroom in our Left Bank hotel, he with his parcel, me fresh from the podium. Without speaking, he unzips my dress. The parcel sits unopened. Dinner can wait.

Bahamas, Winter of 2006

The marriage proposal. At sunset, sitting on a golf cart by a stream, we are scanning the water for the telltale signs of bonefish. I say yes. "Can we marry in the next two weeks?" he asks. "No," I say, with deep regret. "Too much to do with the move to Montreal." It seems like a small detail in the effusive flurry of joy. We head in for cocktails, to celebrate in the Great House at Kamalame Cay. Freeze this moment. I haven't been this happy in years.

Montreal, Winter of 2007

Trudging home from a long day at the office. It's a frigid January evening. I pass many couples heading home together, holding hands,

145

chatting. I pass bistros with twosomes at window tables, communing over menus. It feels like a personal affront.

I pass three small stores selling wine. No, I tell myself: do not stop. I pass a fourth. My willpower fails: I pick up an unremarkable bottle of white. I will not earn a monkey sticker tonight. I will read *Sober for Good*, or maybe *Drinking: A Love Story*. And I *will* have a glass of wine. After three, I will call Jake. I am lonely beyond measure. Nothing is turning out as it was supposed to.

I have a problem, and it's mushrooming fast. My shame is growing, too. There has been an incident, one where I embarrassed myself.

I have started to see an addiction doctor, and I tell him in detail about the event. Too much to drink, blacking out at the end of the evening. I am shattered.

He thinks for a minute. "Did you ever hear of the great basketball star who came back for a season past his prime? He flubbed the game. The crowd booed. Postgame, a reporter stuck a microphone in his face, asking him how he felt. 'I learn—even from the boos.'"

The point? I wonder.

"Those who love you are giving you cues. Learn from the boos."

Georgian Bay, Summer of 2007

Jake and I have vowed not to drink for our entire vacation. For seventeen days in the north woods, we have abstained. Chopping wood, sleeping well, swimming hard: we feel good. We are happy. We have been to my first meeting at a support group, together: holding hands, listening to the stories. I am not certain I am ready to quit forever, but I know I have to tone it down. And we have a new deal: if I ever drink more than two glasses on one occasion, I have to join Alcoholics Anonymous.

Last night, at my own birthday celebration, I had more than two—
many more than two. On waking, I need to gauge the temperature
of Jake's mood, since I can't remember the end of the evening. Jake
is staring at the ceiling. I might as well state the obvious. "I blew
probation," I say. "Yes, you did," says Jake. I am silent. He continues.
"Baby, it's a problem." "I know. I know." I roll the other way, and start
to cry. He continues. "You smell like Brendan Behan." The jig is up.

Winnipeg, Winter of 2008

Jake has a condition called ankylosing spondylitis, something that
was diagnosed at the Mayo Clinic when he was in his teens. One in
ten thousand has it, and Jake was one of the unlucky ones: he's an
adventurer with what he calls a "stiff back." In other words, he can't
turn his head: his spine is fused from its base right up to his skull.
The condition has also meant he has had multiple hip replacements
and now there is a pelvic rebuild—a big operation with a big
recovery, one that I am nursing him through. On day seventeen,
post-op, we manage to make love, gingerly. I am in love. I love being
here, in Winnipeg, together at last. Each night, I cook; we build large
wood fires in the fireplace, light candles, and watch *Prime Suspect*,
with Helen Mirren; we sleep long sleeps in each other's arms.

One night, we watch Mirren—aka Chief Inspector Jane
Tennison—confront her own drinking. Waking in her living room, a
bruise on her head, in silk pajamas, wine and whiskey on the coffee
table. The toilet seat is up in the bathroom. Someone who had
been with her the previous night says: "You remember the pub?
Me leaving?" "No," she admits. "Remember anything?" "No." "Jesus,
Jane, you have to look after yourself."

This we watch in silence. After Jake goes to bed, I press rewind.
Jane at her first AA meeting. A former nemesis is there. She sees

him and heads for the door. He runs after her: "We can't do it on our own, Jane." I press rewind. I know this is my future, but I don't know how in hell's name I am going to get there. "We can't do it on our own, Jane." It echoes in my head.

"Come to bed, baby," calls Jake.

"I will, in a while."

I am too worried for sleep. On a regular basis, I pull out my copy of *Drinking: A Love Story* and re-answer the twenty-six questions: "Have you ever tried to control your drinking by making a change of jobs, or moving to a new location?" Yes. "Have you often failed to keep the promises you have made to yourself about controlling or cutting down your drinking?" All the time.

I stay up late, looking for rehabs in Mexico, Googling places in the southern United States, in the Bahamas. Faraway places, where I can heal out of sight. Jake wakes again: "Come to bed, baby." I crawl in beside him and kiss the back of his precious neck. He takes my arm, holds it close, wrapping it around him. "Don't worry," he says. "We'll figure this out."

The next morning, Jake is holding me. "You remind me of Lucy Westenra, the beautiful woman in *Dracula*. Each night, alcohol comes and steals your spirit, drop by drop." I am speechless. He's right: alcohol is stealing my spirit. "Or maybe alcohol is your pet grizzly bear." Right again: it's bigger than me.

We both know there will be no more drinking: no more sharing a moonlight scotch, no wine with dinner, no champagne at our wedding. Privately, I vow: no wedding at all, until I have nailed this problem, been sober for a year.

How can one write about drinking—or quitting drinking—without addressing wine and romance? Every woman brings it up, sooner or later. "I can give up drinking—but what about champagne at my wedding?"

It's always the champagne at the wedding. Or a dry martini on a date at an elegant bar—maybe the Oak Bar at the Plaza hotel. Romance and the glass: inextricable.

Where did all this begin? For me it started at the movies. Sean Connery, ordering his martini in *Goldfinger*: "Shaken, not stirred." Honor Blackman in the background—the infamous Pussy Galore. I was eleven, and I was smitten. Connery was my first crush. The hair on his chest, the curl of his lips, his Scottish burr, his five o'clock shadow, his crisp white shirts—all of it did me in. On a sleepover, I told Dilly Capstick that I wanted to run away with him, and I meant it, too. I imagined a hot-air balloon, over France, and lots of kissing. She liked the physical details. We stayed up all night devising my plans.

Much later, a beautiful blond would displace Connery—Steve McQueen, to be specific: on his motorcycle in *The Great Escape*. I remember lining up at a pajama party, taking my turn kissing the screen—I kid you not. Later, I would love him swirling brandy in a snifter before a seductive game of chess with Faye Dunaway in *The Thomas Crown Affair*.

My crush on McQueen lasted several years, through my black-and-white phase, when I discovered the pleasures of *The Thin Man* movies, *Breakfast at Tiffany's*, and *Casablanca*—a movie that romances the glass, romances the bar, romances Paris. Of all lovers, no one matches Ingrid Bergman, with her liquid eyes, her perfect skin, her broken heart.

Best memory of all? Being fourteen, watching *Two for the Road*: Audrey Hepburn and Albert Finney, tucked into their little MG, or cuddled up in bed with wine and grapes, feasting on each other, bickering with each other, embracing to a Henry Mancini score. As a young teenager, I could not get enough of this movie—or rather, the dialogue. I renewed Frederic Raphael's script at our small-town library several times in a row. (No one else wanted it.) To this day, I can quote long passages from memory.

But nothing romanticized drinking the way Fitzgerald and Hemingway did. *A Moveable Feast* had an indelible effect on me. For years, *Tender Is the Night* was my favorite novel. At eighteen, I toted Nancy Mitford's *Zelda* and Calvin Tomkins's *Living Well Is the Best Revenge* around campus. Drinking and romance and sophistication and writing—all entwined, with a little madness thrown in for good measure. It was a heady mix.

And then there was the literal romancing of the glass: shopping as a newly engaged twenty-three-year-old, reviewing crystal patterns. Unwrapping gift boxes with care: tall etched flutes, fragile and full of promise. I remember tucking them into the cupboard, knowing they would only emerge on the best occasions in my life, the ones I would never forget. And they did: on birthdays, on anniversaries, on my son's unforgettable christening Sunday. I always romanced the glass.

I am not alone in all of this. At twenty-eight, Julia Ritz Toffoli is the founder of Women Who Whiskey, a Manhattan-based club peopled by eighty-seven young women between the ages of 26 and 32. Many are recent graduate students of Columbia University, where Ritz Toffoli just earned her master's. Now working as a program coordinator with George Soros's Open Society Foundations, Ritz Toffoli loves what she calls New York's speakeasy revival, and considers herself part of a cocktail renaissance. The entrance to her favorite bar—Please Don't Tell, in Greenwich Village—is within a vintage phone booth in a hot dog store. Clover Club, also high on her list, is where Women Who Whiskey celebrated the eightieth anniversary of the repeal of Prohibition, complete with vintage twenties décor and a jazz band. Personally, Ritz Toffoli loves her whiskey neat—either bourbon or Templeton rye—but at heart, she confesses, "I am still a Manhattan girl." This fits: the cocktail is rumored to have been created at the Manhattan Club for a party hosted by Winston Churchill's mother, the Brooklyn-born Jennie Jerome. "The tale has been debunked," says Ritz Toffoli, "but I still like to think it's true, for the time and place it suggests."

Ritz Toffoli comes by her love naturally: her mother is from the Champagne region of France. "Cases of champagne were de rigueur at every holiday and birthday—as an infant, I was even baptized with champagne, holding with a long-standing family tradition. So as long as I can remember, celebration and drinking have been inextricably linked. For me, alcohol is an accessory to joyful events—not an escape from social angst, as it often is for people who started drinking in their teens."

I will never taste a Manhattan, I think somewhat wistfully. This is what I am jotting in my journal as I wait for "Scout," sitting in the Botanist in London, watching her cross the street: a slim, savvy mother of two negotiating her way through the morning crowd, toward this elegant Chelsea bistro. Sitting down, she announces she wants to be known as Scout. "Scout?" I ask. "This is your preferred name? As in *To Kill a Mockingbird*?" (Few women with a drinking problem are willing to be named: the stigma is too large. And if it's a pseudonym they need, I let them choose their own.)

"Yes, Scout," she says. "Best name ever!"

"Harper Lee would be proud," I say.

She grins. "Did you see this morning's news about women and rising liver disease?"

Zestful but clearly tired, Scout has been up half the night. She barely sleeps. "Sometimes I functioned better with a hangover than without sleep," she says with a wry smile, ordering her first of two Americanos.

Scout has been sober—this time—for two and a half years. Like me, she has been a "high-functioning, high-bottom" alcoholic. And like me, she had a wistful feeling about champagne when she considered getting sober—champagne at her wedding, to be specific. For Scout, alcohol was always connected to romance: in fact, she used her very first drink to screw up the courage to kiss a boy: "I could only get it down in a competition—all because I wanted the courage to kiss him. I needed that courage because I felt ugly, fat. I was a very awkward teenager."

Scout was a once-a-week drinker. "Every time I drank, it was a disaster. I would only drink to get drunk. I would never eat—what was the point? It would line my stomach. I loved getting to oblivion: the giddiness, that crazy feeling of the third glass." What triggered her quitting? "It was the blackouts, the shame." Plus a book launch where she had too much to drink and said the following to her then boyfriend: "I would never have sex with you, even if you paid me a million pounds."

"When I first went to Alcoholics Anonymous, I couldn't relate to anything until I read Caroline Knapp's memoir, *Drinking: A Love Story*," she says. "I gave it to my best friend, and she said: 'Did you used to do *that*? Take a swig when I went out of the room?' Of course I did! It was the book that made me accept that I was an alcoholic, that I didn't have to be lying in the gutter to qualify. I was working incredibly hard, had loads of boyfriends, never missed a day's work. I was the lynchpin of my family, with a massive sense of duty. And I was definitely good fun—for the first ten years. I was quite shocked when I discovered that I had developed a problem with boring old alcohol."

Scout went to Alcoholics Anonymous for five and a half years. She didn't get a sponsor or do the steps, but she stayed sober. She remembers this time as one of the happiest of her life. "Real contentment," she says. "Calm."

Then she met her future husband—an active alcoholic. Within six days, they were engaged. They had a picture-book three-day wedding in France. "All my life, I imagined my wedding day with champagne. On the day, I didn't even think about it. It was the last thing on my mind."

Within weeks she was pregnant. "After my son was born, I thought my husband didn't fancy me anymore," says Scout. "I very consciously got pregnant again—I didn't want my son to be an only child." But after her daughter was born, "it was clear that the marriage was going wrong. My husband was an active addict. I wanted to connect with him. It started with an innocent glass of champagne. I remember being at an event and someone said, 'Come on! Have one glass of champagne!' I had one glass. Then one glass of wine. Then I decided to drink only wine. Soon I was blacking out on a regular basis. And this time—just as they tell you—it was a much quicker decline to rock bottom. It wasn't long before I was in a real pickle."

Scout's husband had an affair and the marriage soon ended. "I lost my job and my marriage all in the same year. I got a new job, and my boss would come up to me and say, 'Do you want to know what you said to me last night?'

"I had two children, but all I wanted to do was to have a breakdown: go to rehab. In my last five years of drinking, I really craved that feeling of peace. My brother asked me what I needed. Therapy? No, I need to go back to AA. I have always wanted a manual to life. If I do AA, it keeps me tethered to the ground."

Today, none of her family knows she goes to meetings, although she believes all of her siblings are genetically predisposed to the disease. She is fully accepting that she will never have another drink. "I don't suit drinking," she says. "Nor do I suit tomatoes. It's as simple as that." Does she miss it? "Sometimes," she says, with a wistful grin. "But only at really happy times: sitting on a balcony, looking at a view."

Like Scout, I occasionally miss drinking at the happiest of times. I remember an awkward moment on my first sober birthday. Arriving at my favorite restaurant with my handsome twenty-three-year-old son, I was confronted with a difficult situation. The owner, who knew me all too well, delivered slim stems of champagne as we sat down, a red raspberry floating in each glass. I was only four months out of rehab. He might as well have placed a nuclear bomb on our table. Both Nicholas and I froze. "Get him to take them away," whispered my son. "Let's just give it a minute," I whispered back, reluctant to cause a scene. With that my son stood up and exited the restaurant. It took twenty awkward minutes and a few tears before the evening resumed. There's a beautiful photo of us at our table that night, one I still can't look at without wincing.

Four years later, I'm more seasoned in handling these moments. But certain occasions are still tough. New Year's Eve, always. Each December 31, a group of us dress to the nines, gather in a home, and feast on oysters and lobsters. Each year I turn down the champagne that greets us on arrival—seductively blond, bubbly, and beautiful, the perfect accessory to a black-tie evening.

A small handful of us are sober: two men and myself. The men take "smoking breaks" on the back porch in their tuxes. I don't. I drink mineral water at midnight. To this day I am a little wistful as the countdown begins. Always the designated driver, the presenter of the post-midnight cheese course. When the couples kiss, holding their bubbly aside, a little part of me still crumples up like it did in high school, when no one asked me to dance.

What role has alcohol played in my romances? A large one. I know I had too much to drink the night I met Jake. Would I have had the courage to invite him home without being a little tipsy? Doubtful. No question, a little wine helped smooth the way for that first encounter. Ditto the one with Will. And, God knows, I am not alone.

I decide to ask a woman who dates a lot what she thinks of the relationship between alcohol and romance. Alexandra is an actress and a jazz singer, exquisitely sculpted from her high-set cheekbones to her polished toes. Men stare at her, women envy her, and she would have it no other way. This is her business, and she works hard at it. She is a single woman, keen on men and dating.

The daughter of a mechanic and a hairdresser, Alexandra is deeply aware of growing up on the wrong side of the tracks. Her first memory of alcohol is from a sleepover at a wealthy girl's home: she saw their wine cellar, and never forgot it. She found alcohol in her twenties, starting with gin and tonic in between sets as a singer. "Suddenly I felt like one of those rich girls," she says. "It was like a childhood dream realized. The flip-flops were gone and I was wearing high heels. It made me feel wonderful."

Soon she was engaged, and her fiancé introduced her to fine wine. "The whole romance was alcohol-fueled," she says. Before long, however, they began to fight. Alexandra determined it was the red wine. "Suddenly, you go gremlin on me," her fiancé would say. For a short while she gave up wine. But soon the rule was broken: the two traveled with his family to a wedding in France, and she drank on the overseas flight. "It was France, for goodness' sakes," she says. "And I had never been to Europe." There she mixed spirits, champagne, and white wine. Soon she was "puking my guts out. I used to put a bucket by my bed." On one memorable evening, she bad-mouthed her fiancé to his entire family, itemizing his faults. "Alcohol lobotomizes you. I don't really remember everything I said, but I knew it was over after that. There's only so much you can do before you cross a line."

With that she moved out, and joined AA. "I was single, and sober, and it was great. I was in a TV series, made a lot of money, and I was doing really well. I forgot all the hell of drinking. I felt cleansed: the past was over."

Alexandra fell in love with a wealthy, older, married man. "I remember the first drink—he supported it," she says. "I thought: 'What's the worst that can happen?'" He bought expensive clothes for her, and took her to London. "It was the time of my life, truly magic. He made me feel like a princess. There were nights when I got completely drunk, but so did he. I was just so happy to be there. I thought I had arrived."

Looking back, Alexandra sees that she was "sort of a high-class prostitute, a professional call girl." Each time her man visited, he would bring a case of wine. "I would drink two more bottles alone, after he left. I would resent that I was always alone on Christmas, on weekends, on my birthday.

"The beast was back. I stopped going to auditions. The worst happened at one of the most beautiful restaurants in town: I don't remember eating. I vaguely remember him yelling at me for my behavior before he dropped me off. I then went into a very dark period where I was drinking three bottles a night—buying one and then getting in my car, drunk, to buy the next two.

"And so I got sober—it was either that or kill myself. I ended up doing the long, slow journey back. I miss alcohol when I see social drinkers—a couple drinking cold glasses of wine on a patio. I miss when it was good. But I don't miss the end. There was never enough. From the time they pulled the cork until they poured the first glass, I would get so angry at waiters: 'Hurry, will you?'"

Today Alexandra asks herself: "Would I be willing to risk my whole life on a social beverage? And the answer is no. But you know what they say about alcoholics? The alcoholic reaches the end of her days and St. Peter says: 'Heaven or hell.' Only the alcoholic says: 'Can I see what hell looks like first?'"

For me, the incident that ruined my romance with drinking involved not Jake, but my son, Nicholas. It was the spring of 2005, and I had a root canal—one that went badly. I was put on heavy antibiotics and told not to drink for three weeks. Near the end of that period, I received a Mother's Day card from my twenty-year-old son, a hand-made card titled "Happy Mother." There I am, at a typewriter. Note, it says, "the whites of her eyes are white." Also note: "She is drinking Perrier, not wine."

That card walloped me where it hurt, and it ended my romance with the glass. For three years I carried it everywhere, tucking it into my diary or my daybook: the truth writ large from a man I loved, one I had let down and who was brave enough to tell me how he felt. This took guts.

Long before Jake said anything, or my sister Cate or my friend Gillian, Nicholas's eloquent drawing ruined my love affair with drinking. True, it took me another three years to quit. But never again was it easy to see the experience as innocent, as something untainted by trouble. For me, the romance was over, and the tough part had only just begun.

10.

The Modern Woman's Steroid

POPPING THE CORK ON MOTHER'S LITTLE HELPER

I can bring home the bacon, fry it up in a pan,
and never let you forget that you're a man.

—ENJOLI PERFUME AD, 1980

("THE EIGHT-HOUR PERFUME FOR THE TWENTY-FOUR-HOUR WOMAN")

Is alcohol the modern woman's steroid, enabling her to do the heavy lifting involved in a complex, demanding world? Is it the escape valve women need, in the midst of a major social revolution still unfolding?

For many women, the answer is a resounding yes.

Racing in from a long day at the office, an evening of cooking and homework ahead: the first instinct is to go to the fridge or the cupboard and pop a cork, soothing the transition from day to night with a glass of white or red. Chopping, dicing, sipping: it's a common modern ritual.

For years it was me at the cutting board, a glass of chilled white at my side. And for years this habit was harmless—or it seemed that way. My house wine was Santa Margherita, a pale straw-blond Italian

Pinot Grigio. There was always a bottle in my fridge, and I'd often pour a second glass before dinner, with seeming impunity.

In the years when this was my routine, I rarely thought to put the kettle on instead. These days, my go-to drink is Celestial Seasonings Bengal Spice tea: a rich mix of cardamom, cloves, chicory, cinnamon, pepper, and ginger. But back then, as I burst through the front door, laden with groceries, wound up from the day, my first instinct was to shed some stress as quickly as I shed my coat. Once, after an unusually difficult day, Jake pointed out that the fridge was open before my coat was off. It pained me to hear this, but I know it was true.

Within a few minutes, I would be standing at the cutting board, phone cradled on my shoulder while I sipped and chopped and chatted, often to my friend Judith or my sister, Cate. Nicholas would be upstairs, doing homework, and dinner would be in process. Sip, chop, sip, chat, exhale, relax. Breathe. With two parents who had their own serious troubles with alcohol, alarm bells should have been ringing. But my habit seemed relatively harmless. Common, even. A glass or two seemed innocent enough.

And truth was, believe it or not: I got a lot done when I was drinking. In my alpha dog years—when I was holding down a senior job at a magazine, raising an artistic, athletic young man, giving speeches on the circuit—life was more than full. Alcohol smoothed the switch from one role to the other. It seemed to make life purr. I could juggle a lot. Until, of course, I couldn't.

That's the thing about a drinking problem: it's progressive. But for a long, long time, alcohol can step in as your able partner, providing welcome support—before you want to boot it out.

On a recent November evening, I took a stroll through the elegant streets of London's Chelsea district around that witching hour—an hour when many had yet to pull the shades for the evening. Heading up from the Thames River, north on Tite Street, I passed more than one window with a woman standing at her kitchen counter, a half-drunk glass at her side while she worked on the evening meal. I passed a dad unloading children from a shiny BMW, children lugging heavy knapsacks, calling out to younger siblings waving in an upper window.

It was a cozy scene, and I found myself thinking wistfully of those rituals of younger years, when my son was under my roof—not far away in California, doing a master's degree in fine art. Time was he would saunter into the kitchen, hungry and tall, and dance me around the room while dinner cooked—a boisterous little tango that left me flushed and laughing. More often he would serenade me with his guitar.

Those years were loud and rambunctious and incredibly busy, crammed with duties and chores. Once dinner was over, he'd do homework and I'd make lunches and then noodle with a little more work before bed. He was a rower and morning came early: I'd rise in the dark and ferry him down to the waterfront, standing with the other parents as the boys headed out on the water.

Those years were full of stress and laughter, in equal doses. Often, Nicholas and I would find ourselves up at night, talking in the kitchen: I would make popcorn and we would stand side by side, filling in the blanks for each other. We were a pack of two: our conversations were deep and rewarding, and we read each other easily. And when those precious years were over, when he went off to university, the house became very quiet. Too quiet: like a stage set after the actors exited. That's when I wrote a column in the magazine, called "Mother Interrupted." And that's when I began to think that a third drink might make sense. And once it was three, I was in trouble.

Flying over to Britain, to do research for this book, I splurged with my airline points and booked myself a first-class ticket. Flight attendant to me, after dinner: "Would you care for some port with your cheese, madam?" "No, thank you, I have to work." She frowns. "Lots of people drink port while they work." And indeed, she pours some for the neighboring woman, who is laboring over a spreadsheet with a glass of wine. All I can think is: "That used to be me." Six years ago, that would have been me, and my exit from the plane would have been a little fuzzy.

In a recent poll done by Netmums in Britain, 81 percent of those who drank above the safe drinking guidelines said they did so "to wind down from a stressful day." And 86 percent said they felt they should drink less. Jungian analyst Jan Bauer, author of *Alcoholism and Women: The Background and the Psychology*, believes women are looking for what she calls "oblivion drinking." "Alcohol offers a time out from doing it all—'Take me out of my perfectionism.' Superwoman is a cliché now, but it is extremely dangerous. I've seen such a perversion of feminism, where everything becomes work: raising children, reading all the books, not listening to their instincts. The main question is: what self are they trying to turn off? These women have climbed so high that when they fall, they crash—and alcohol's a perfect way to crash."

I ask Leslie Buckley, the psychiatrist who heads the women's addiction program at Toronto's University Health Network, if she sees a pattern in the professional women who come to see her. She doesn't skip a beat: "Perfectionism."

Such an unforgiving word, such an unforgiving way of being—echoed by yet another doctor, who speaks of patients who look like they stepped out of *Vogue*: perfect-looking women with perfect children at the right schools, living in perfect houses, aiming for a perfect performance at work, with eating disorders and serious substance abuse issues.

The tyrannical myth of perfection: it seizes the psyche and doesn't let go. My mother was in its grip, and she paid a serious price for it.

This was in the 1960s, when men came home from work and expected dinner and a stiff drink—except my father was usually traveling. For years my mother held down the fort. She wrote perfect thank-you notes, she cooked perfect meals. As a new bride, she ironed bedsheets and pillowcases; as a new mother, she starched our smocked dresses. My sister and I wore white gloves when we traveled, velvet hairbands in our hair, and wrote perfect thank-you notes, too. And then my mother was the one with the stiff drink, and it all crashed—but not before I had it imprinted on me: perfect was the way to be.

Perfect has been the way to be for several generations of women. I don't remember my grandmothers suffering from this syndrome: women who raised families during the Depression, who baked and gardened and read well; who were fundamentally happy, and felt no pressure to look like stick figures.

But those *Mad Men* years took their toll. My mother wasn't the only one self-medicating with a combination of alcohol and a benzodiazepine called Valium. By the end of the sixties, two-thirds of the users of psychoactive drugs—Valium, Librium—were women. In fact, between 1969 and 1982, Valium became the most commonly prescribed drug in the United States. In 1978, it was estimated that a fifth of American women were taking "mother's little helper," as the Rolling Stones called it.

By that time, its addictive properties were well known—and if they weren't, the 1979 bestselling memoir *I'm Dancing as Fast as I Can*, by Emmy Award–winning Manhattan producer Barbara Gordon, blew the lid off. My mother weaned herself from the drug, with the help of rehab, and emerged a somewhat reformed perfectionist.

It never occurred to me—not for years—that alcohol was the mother's little helper of my generation. But it is.

Today, women arrive home from work to face more work. So too do men—but there's a difference. My ex-husband, and the man with whom I shared Nicholas's rearing, is not a perfectionist. Constant? Always. An excellent father? The best. But I never considered him accountable in the way I was for certain essentials. We had a division of labor that worked well: he coached the sports teams, taught our son to ski, oversaw math. When it was Nicholas's turn to eat at Will's, there were three options for dinner: Kraft Dinner, Lean Cuisine, or take-out chili. It never varied. Dinner at my house was more nutritious—but often late. Breakfast was pancakes, from scratch. True, this brought me joy. So did making the Halloween costumes. I was not willing to miss out on some of the essential pleasures of being a mother just because I worked. And I wasn't willing to miss out on some of the essential rewards of a great career just because I was a mother. As a result, my life was complex, truly jam-packed like a Christmas cake. If I could stuff in one more cherry, I did.

Truth be told, Will helped me do so: he did a lot of the ferrying of boys to and from events, up to the cottage for winter weekends. But I clung to the more traditional division of labor, and dined out on stories that bolstered my position. Like the time I came downstairs as a new mother, having allowed myself to sleep in. Will was reading the newspaper in the kitchen, arms wide, Nicholas at his feet. "How's our boy?" I asked. "Just fine," he said. "He's right here." With that, I saw my son, in yellow fuzzy sleepers, look up from the dog dish, a mouth full of kibble.

I had surgery when Nicholas was two months old. Will handled our newborn when I was in the hospital. When I got home, I asked a classic new-mother question: how did you manage the shopping, with the baby in the car seat, in the cart? "Oh, that's not how you do it," said Will. "You leave the cart at the end of one aisle, grab a few groceries, and then return to check on the baby." "And what if someone decides to steal him while you're shopping?" I asked. He didn't have an answer.

These stories were anomalies, but the truth was, I always wanted to be the alpha dog when it came to our son. From the time he was born, I felt that Nicholas was an egg I carried on a spoon, one I was not to drop. I'm sure Will felt no differently, especially as the years wore on and Nicholas evolved.

For my own reasons, I spent a lot of time experimenting with my own customized formula of work-and-home-life balance. I experimented with part-time, flextime, and a journalism fellowship that sent me back to school when Nicholas was two. I tried it all. And when my marriage of twelve years collapsed, I quit my job of twelve years at the same time: I stayed home for the next eighteen months, using my savings to make ends meet. I figured that just as my son had lost, so too would he gain. Ending my marriage was extraordinarily painful, and that eighteen-month immersion in motherhood was necessary and healing.

Once that period was over, I was back to work full-time, with gusto. My son was seven, and I couldn't afford a nanny. I shared some after-school babysitting and took on a project that became one of the most successful in Canadian publishing, winning a National Magazine Award that first year. It was a fifty-page examination of higher education, featuring rankings of Canadian universities. The magazine "went to bed" on Halloween: I made the costume, but I wasn't out trick-or-treating with my tiny knight that evening. Will was.

Lean in, lean back: I've done both, sequentially. I've sat at home, in tears, believing I would never enter the workforce again. And I have sat at the office, exhausted, knowing I was missing a precious evening at home. Both positions have their downsides and their sweet rewards. One thing is for certain: straddling both roles can turn you into human Silly Putty. I remember when my son was born, receiving a card from the writer Marni Jackson—author of *The Mother Zone*—who wrote, perceptively: "Welcome to permanent ambivalence."

"How do you juggle it all?" As Tina Fey wrote in *Bossypants*, it's the rudest question you can ask a woman—ruder than "When you and your twin sister are alone with Mr. Hefner, do you have to pretend to be lesbians?" . . . "You're fucking it *all* up, aren't you? their eyes say."

There were times when I did mess up. One winter, Nicholas came down with a bad case of whooping cough. (Turns out he and his pals had decided snow jackets were for sissies, playing every recess in their T-shirts.) I spent many nights awake, in his room. One morning I slept through the alarm. This happened to be the day the publisher of Mc-Clelland & Stewart was coming to the editor's office to discuss a possible book contract—one I was to oversee. I missed the beginning of the meeting, but the publisher was gracious. He stood and shook my hand, and said, "Hats off to mothers." You don't forget a moment like that.

It was twenty-one years ago when I returned to work, full-time—the same year Hillary Clinton defended her personal choice with the following: "I suppose I could have stayed home and baked cookies and had teas, but what I decided to do was fulfill my profession." At the time, her comment drew scorn from many, but I was cheering. It was a pivotal moment in the mommy wars: the tension was deep.

Of course, this was also the era of Martha Stewart, who had a decade-plus run as the queen of perfectionism until she was incarcerated. Homemade Christmas ornaments were all the rage, and Martha was dictating the rules. Here's a slice of her December to-do list, published helpfully at the front of *Martha Stewart Living*: by December 8, all fruitcake baked; by December 10, all gingerbread houses assembled; clean chandeliers on December 11. And so on. Women were outdoing themselves at work and on the home front, contorting themselves like Gumby in the process. Each year, like so many others, I performed the Christmas triathlon, and ended up sick or tired or both. After a few Sisyphean seasons, most of us realized that the more we outdid ourselves, the more we were undone. I cried uncle.

As the late Laurie Colwin once wrote, "It is my opinion that Norman Rockwell and his ilk have done more to make already anxious people feel guilty than anyone else." It was up to us, she said, to reinvent traditions to make way for what she called life's one great luxury: time together.

I took her advice seriously and tried to make room for that luxury. Many of us did. As life continued to speed up, especially with the introduction of smartphones, the need to slow down *fast* became increasingly attractive. In the 1990s came the proliferation of wine bars. In 2000, Time Inc. launched *Real Simple* magazine. In 2004, Carl Honoré's *In Praise of Slow* hit the bestseller list. (Nice thought, but somewhat beside the point when you had carpal tunnel syndrome from overworking your BlackBerry.)

Long before that, I was using wine to decompress, to ease into the second shift of the evening—and so too were my friends, both the stay-at-home mothers and my professional peers. As many women discovered, a drink is a punctuation mark of sorts, between day and night. "It's a shift of gears," says Janice Lindsay, author of *All About Colour*, and mother of two grown children who have both returned home. "A glass of water doesn't make me feel spoiled. A glass of wine says, 'Now you can enter the pleasure part of your day.' I put on some music, and it's a treat, even if I'm chopping onions. What else can we do? A massage is almost a hundred dollars and it takes an hour I don't have. Wine is right here, right now, and I can share it with whoever's with me."

Danielle Perron, the only female partner in a Toronto marketing company and the married mother of a five-year-old boy, pours her first of three glasses of white wine every night at six o'clock. She drinks that first one with dinner. She has the second after her son is in bed, and her third at ten o'clock. Each and every night, like clockwork. "I drink until I'm comfortably numb," she says, "with the perfect sleepy buzz. I'm groggy almost every morning. I get my ass out of bed and I go for a run. Life is high stress, and I juggle a ton of balls every day. This is about peace, the right glass, a ritual." Would she call herself an alcoholic? "No, my dad was a hard-core *Leaving Las Vegas* alcoholic—before he quit forty-three years ago. Never would I call myself an alcoholic. But am I dependent? Yes. For me, that glass of wine is a total joy."

A total joy that is causing her grave doubts. The day I interview Perron, she is on her third day of a twenty-one-day cleanse, eliminating wheat, sugar, and wine. "I don't like that I drink every day," she says. "For the last year, I have been questioning it more. If I can do this for twenty-one days, I will give myself permission to continue. And if not? If after this, I'm still jonesing for a drink at six o'clock? Well . . ." Her voice trails off. She seems uncertain of the answer. "There are so many moments in life that are all about having a glass of wine. And if my drinking time is impeded—if I'm at a play, or working late—I feel aggressive. My glass of wine will be full, and gone in two minutes."

Her friend Paige Cowan, with whom she is sharing the cleanse regimen, is clearer on the outcome. Cowan is a tall, expressive woman who owns Wild Bird, an eclectic little store in midtown Toronto that sells a wide variety of seed for birds and food for the spirit as well: books on Buddhism, meditation, healing. On a snowy winter day, I find myself drinking delicious coffee in her airy living room, nestled on citrus-colored armchairs, listening to her story of wrestling with alcohol, and the role it has played in her life. "In my twenties, it was just about having fun—it was so normalized to drink at the cottage. Partying was really a rite of passage. Then I had my two boys when I was twenty-seven and twenty-eight—that's when I began to be conscious of my patterns with alcohol."

Growing up, Cowan found herself without much parenting: her

mother was a serious "self-medicator," with pills. "This time last year, I became mindful about my relationship with alcohol," she says. "Was I having a drink to deal with anxiety, self-medicating?" She decided to give up alcohol altogether—and not because she was an alcoholic. That's when the pushback started. "Pretty much everyone I know is heavily into alcohol," she says. "They disguise it as something sophisticated or chic. It's uncomfortable when you don't drink. People ask: 'Have you stopped drinking altogether?' Not everyone, but most. But I have noticed a big difference—and so has my husband. I have more vibrancy, my sense of humor is back. Alcohol adds a cloud, and the cloud lifts. It makes you wonder: 'What was I doing to my body?'"

Perfectionism is a culprit that Cowan knows all too well. "At one point in my life I was trying to be the perfect woman: doing things in the community," she says. "For a good ten years, I was unconsciously driving my life—and that's when I self-medicated the most with wine. I was involved in so many community efforts—it was that feeling that I was never good enough. That whole perfectionist thing was driving everyone: you could bust your ass, and it wasn't good enough. A relentless standard of perfection. I found it shocking how hard women are on other women. At our little school in a pretty little neighborhood, there was an abusive standard of perfection. You would often hear women say, 'I'm going home and having a glass of wine'—as a release."

"This is the way we are," says Cowan. "We encourage young women to live their lives a certain way—and it has nothing to do with what feels right. We tell them they're not pretty enough: that's what we bombard them with. Get on the treadmill, bust your ass at work. I think we're living in a culture that's so demanding: you never feel like you're good enough. It wears people down. People are exhausted at the end of the day. They go home and have a drink as a way to cope with all of this—a lot of people have to self-medicate because it would be hard for them to look in the mirror otherwise. The whole concept of being conscious—that's hard work. A lot of people just don't want to sign up for it."

Signing up to be conscious: this is what Lisa decided to do after many years of drinking too much. A prominent woman in her sixties with a packed Rolodex and a full calendar, Lisa is a mover and shaker. The former senior manager of several companies, she has spent much of her life in her adopted home of Canada, but has now settled back in the States near her grandchildren, in Chicago. Raised in Cleveland in an upper-class home, she says bluntly: "I had two raging alcoholic parents—a rich couple who lived a crazy life. They had a house in Florida and belonged to a club where there were many so-called functioning alcoholics—on the tennis court at seven a.m., drunk at lunch, asleep all afternoon, drinking again at five in the afternoon, in bed at nine. Repeat.

"My father drank himself into bankruptcy. He would scream and yell and leave the house in the middle of the night. Me? I didn't stand a chance. I have two brothers—one who has been in and out of Hazelden and the other who is in AA.

"I actually remember the first time I got drunk," says Lisa. "It was at Brown University, at a football game. I was visiting a boyfriend and I went back to my little hotel room and passed out. I was so horrified and ashamed, I didn't answer the door when the guy visited that night."

Lisa spent the 1970s in New York, working as a political lobbyist. "I would call people completely drunk," she remembers. "I was conscious I drank too much, and I didn't want to face it. I remember a birthday party for Bella Abzug on the top of the World Trade Center: I got blotto and ended up going home in a cab with a friend who called me on it—but I didn't want to hear it."

Married, she and her husband, Henry, ended up moving to Toronto when their three children were small. "It just got worse," says Lisa. "For the first time, I wasn't working, and I invited a group of women over for coffee. I remember going to the kitchen and filling my coffee cup with vodka. When they asked for my phone number, I couldn't even write. Needless to say, I never heard from them again. But that didn't stop me from drinking too much in front of my husband's boss, and embarrassing the hell out of myself at a dinner party."

Was she efficient when she drank? "Yes," says Lisa. "I would pour a drink and stay up half the night to get stuff done, whether it was organizing Halloween costumes or pulling recipes for a dinner party we were having. All of those years, I would just run from one thing to the next, not really thinking. I was such a boozer—but I was an amazing organizer when I drank. Partly, it was to prove that I wasn't a drunk; part of it was compensating for my drinking, my lying about not drinking. I was up early, making breakfasts. I used to go to McDonald's every day on the way to work, ordering a hamburger—all that grease seemed to help with the hangovers. On the way home, I would say: 'I will only have one drink tonight' . . . and I always failed."

She's the first to admit she had a couple of close calls with her children. "One Sunday, we had friends over for brunch, and drank champagne. After they left, I decided to make hamburgers, while I had another glass. I started cooking them under the broiler and there was a fire. Henry raced in and he knew I was drunk. I also drove drunk once with the kids in the car. I was just sloshed and went straight to my bed when we got home, passing out. My eldest woke me up because he had been sick—and I had no awareness whatsoever. I was so lucky that I never set fire to the house or killed anybody."

One night, sitting alone with a scotch, Lisa was watching TV and Betty Ford came on. "She said there are two things you need to know: One, it's the first drink that gets you drunk. Two, this is a progressive disease—it only gets worse. That was the moment for me:

if you start drinking and can't stop, you're an alcoholic." Several days later, dressed in her Max Mara suit and her pearls, Lisa headed to AA. "The minute people started talking, I realized: 'This is my life.' I haven't had a drink since. I may not have hit a horrible bottom, but I could see it, and it was terrifying."

Lisa isn't alone in pushing her drinking too far. Jennifer, who worked in sales, quit as well—but not before her drinking helped her get ahead in business. "Alcohol made me social—it was a lubricant, and allowed me to be more gregarious," says the wealthy sixty-year-old, now retired. "I would end up, at the end of a conference or a show, in the bar with the guys to the wee hours, learning a lot from key businessmen. They were my mentors. We would go to a restaurant and after dinner, they would order shots of Kahlúa. I wanted to learn, so I drank what I could to keep up. Did it enable me to work harder? Yes, it enabled me to keep up the pace."

And as we all know, keeping up the pace is everything.

As Jennifer makes clear, prosperity has presented options that didn't exist for other generations. Professional women join their male counterparts after work, going drink for drink. Going drink for drink can be problematic, to say the least. As high-profile Toronto addiction counselor Andrew Galloway says: "In a work situation, who has the guts to say, 'We'll have another round—but leave her out of it'?"

What Galloway sees is a new generation of successful women in their late thirties to mid-forties, heading to rehab. Usually they choose high-end facilities, charging in the tens of thousands. And the biggest news? Like me and like Jennifer, they have the resources to pay for it themselves.

Perfectionism, alive and well, on a whole new level.

11.

The Last Taboo

DRINKING AND PREGNANCY

This may be the most stigmatized area
of a very stigmatized subject.
—JANET CHRISTIE

Is it safe to drink while you're pregnant? When I was pregnant with Nicholas, the commonly accepted answer was no. But in recent years, with equal certainty, science has given women three definitive answers: no, yes, and maybe.

In 2010, a widely reported British study stated that the children of mothers who drank small amounts of alcohol during their pregnancy were not at an increased risk for behavioral or intellectual developmental problems. The study, which was published in the *Journal of Epidemiology and Community Health*, went further, saying that children of light drinkers were 30 percent less likely to have behavioral problems than children whose mothers were abstinent during pregnancy. The study also reported that children of light drinkers achieved higher cognitive scores than those whose mothers had abstained.

I remember being outraged by the study, and its success in

undermining years of messaging on the risk of alcohol to the developing fetus. I wasn't alone. But the comments sections of newspapers were filled with this sort of response: "My mother drank when she was pregnant with me, and I turned out just fine. I intend to do the same thing." To which others responded: "Yes, look how you turned out: irresponsible and brain-damaged." It went on and on: page after page of jousting. Clearly, this hit a nerve.

I turn to Sterling Clarren, CEO of the Canada Northwest Fetal Alcohol Spectrum Disorder Research Network and one of the world's leading researchers on fetal alcohol spectrum disorders (FASDs). He is measured in his reaction: "The U.K. study was unfortunate. Saying that you can have one or two is too simple."

What if a woman drank before she knew she was pregnant: say, a glass of champagne on her birthday? "Obstetricians are confronted with this kind of question all the time," says Clarren. "Drilling down is important. Was it really just one glass or half a bottle? How big was the glass? I have three kinds of champagne glasses at home." He continues. "Was that really the only time you drank? Are you five-nine and heavy, or a small woman? The only ones who really know exactly what they drank are the beer drinkers because today, no one drinks standard drinks. Precision in drinking? That's not how the world works."

In the end, says Clarren, "all we can tell a woman is whether she is at high, medium, or low risk, and she can make it lower if she doesn't drink anymore. The reality is, there is a relative risk to drinking."

What if the woman's answer is different? "Voluminous amounts in early pregnancy once or twice a week? That doesn't translate to one hundred percent risk. Fifty percent risk is more likely. Women want to know what low risk is. They're asking for a simple, fair discussion of this."

According to Clarren, there is a complicated formula for dose and effect, involving how much was consumed, the timing in pregnancy, and a host of other factors: the mother's genes, the fetus's genes, whether the mother smokes, her potential vitamin deficiencies, and so on. "It's a complicated formula, and we don't know how to fill it in—I doubt we ever will. And when we talk about risk, the question is: Risk for what? For massive malformations of the brain? For blindness? For diminished executive functioning? A mild brain disorder? No one knows the answer."

So, what about that glass of wine on your birthday? "Of course, you can have it. You're taking a risk—a very small risk. A true small amount on a rare occasion is not very risky." He pauses. "But I don't know what rare is, and I don't know what small is. How much mercury is safe for the fetus? Raw cheese? We just say avoid it. The advice is the same with alcohol because we just don't know."

Canadian expert Nancy Poole, well known for her collaborative work on FASD-related research and her work with women, acknowledges that many want black-and-white answers. "Unfortunately," says Poole, "the territory is gray. To represent the risk accurately for a wide range of women, I like to say simply that it's safest not to drink in pregnancy. We need to balance the knowledge that alcohol is a teratogen and that one drink is unlikely to cause harm."

Poole believes that there is a stumbling block in doctors' offices: "When you scratch the surface, yes, they are asking the question— 'You're not drinking, are you?' But many physicians don't know what to do when the answer is 'yes.'"

Obviously, there is great stigma around drinking and pregnancy. Many women with a substance use issue are afraid to admit to having a problem, for fear their child or children will be apprehended by a child welfare agency. Should they want help, there are challenges: most treatment centers do not accommodate children, or modify programs for pregnant women.

Meanwhile, with a larger number of women drinking at risky levels, many developed countries are facing an epidemiological perfect storm. Sixty-two percent of the babies in Canada are born to women between the ages of 25 and 34—the group demonstrating the fastest growth in risky drinking. The next-youngest age range is "drinking off the charts," according to Gerald Thomas, a senior researcher with the Canadian Centre on Substance Abuse, located in Ottawa. "They're living the culture, drinking hard." These women may quit drinking once they know they are pregnant, but a great deal of damage can happen in the weeks before pregnancy is confirmed. Says Thomas, "The costs associated with FAS—a lifetime disability—are massive."

It has been roughly forty years since fetal alcohol syndrome (FAS) was first described as a clinical diagnosis, and it is widely accepted as the leading developmental disorder in the world. It is the primary cause of mental deficiencies in developed countries. Fetal alcohol spectrum disorder, or FASD, is an umbrella term that encompasses four lifelong conditions including: FAS, partial FAS, alcohol-related neurodevelopmental disorder (ARND), and alcohol-related birth defects (ARBD). All are preventable and are caused by prenatal exposure to alcohol. FASD is associated with a broad array of physical defects; cognitive, behavioral, emotional, and adaptive functioning deficits; congenital anomalies, such as malformations of the cardiac, skeletal, renal, ocular, and auditory systems, among others.

Full-blown FAS and partial FAS—typically, one case of the first for every four cases of the second—are marked by a set of physical conditions, with partial FAS having a subset of those fully observed in FAS. These include a flattening of the middle groove between the nose and the upper lip, known as the philtrum; a thin upper lip; small eye openings; droopy eyelids; a wide distance between the two inner eyelids; a small head. The more severely affected the facial features, the more severe the brain effects.

Alcohol-related birth defects relates to the damaging teratogenic effects that alcohol can have on developing cells and organs. Alcohol-related neurodevelopmental disorder is typically referred to as an invisible disability without the distinctive dysmorphic face and classic growth deficiency. It is represented by a complex pattern of behavioral or cognitive abnormalities.

The consequences of FASD are diverse, affecting individuals, their families, and their communities in a broad way. People with FASD may have significant issues with memory, attention, decision making, and self-care. They may have problems with organization, completing tasks, controlling their emotions. These impairments are also accompanied by complex and highly detrimental health conditions. Without crucial support, those affected by FASD are at high risk of developing such secondary disabilities as mental health problems or alcohol or drug issues. They may have difficulty with the law, school, employment, and homelessness.

"There is no amount of alcohol that has been proven to be safe during pregnancy," says Svetlana Popova, a senior scientist at Toronto's Centre for Addiction and Mental Health and one of three leading investigators on a twelve-country international study on the prevalence of FASD. "Even half a glass of wine can be damaging to a fetus. It could damage any one of a number of organs, or the central nervous system."

Twenty-two years ago, Lynn Cunningham's stepdaughter gave birth to Andrew, a child with FAS. When he was eighteen months old, Cunningham and her husband took the boy into their home. "He was the child who would not turn off," says the Toronto writer and professor. "Andrew was indefatigable, and would bounce around, creating messes wherever he went. He would do things that even little kids knew not to do—jump in puddles near laundry. All the clothes were marked by mud. It didn't extend to doing extremely dangerous things—he wasn't a darter into traffic—but he didn't have a sense of caution. He used to fly into rages and rants." She called him "Tornado Andrew."

Today he is enrolled in a music production course at community college—an extraordinary success story for someone born with FAS. He's living on his own. He's not in jail. Still, he counts on Cunningham for a great deal, including a weekday wake-up call. According to Cunningham, conventional wisdom says that those with the syndrome operate at two-thirds their chronological age. That will put her close to eighty by the time he will be functioning as a twenty-five-year-old. "The issue is: he relies on me for a lot of stuff—am I going to be able to do this for him as long as he needs me?"

Common logic says that the worries troubling Cunningham will be shared by a growing number of women, if you consider the growth in risky drinking. As she wrote in a magazine memoir: "The perception that FAS is largely confined to the indigenous population is an enduring fallacy."

Jan Lutke agrees. The Vancouver mother has adopted twenty-five children, sixteen with the diagnosis, ten of whom still live with her. She is well seasoned in the disability and its many ramifications. Says Lutke: "If you're white, you're ADHD. If you're not, you're FASD. If you look at an aboriginal kid, it's the first thing people think of. If you look at a Caucasian kid, it's the *last* thing people think of—as if skin color protects you from the effects of alcohol. We think this disability only affects the disenfranchised or the poor. This is not an 'us-or-them' issue: it's an 'us' issue."

Looking at the results of three studies, one might conclude that the older, educated, professional woman is more likely to drink during her pregnancy than her younger, less-educated counterparts. Perhaps not surprisingly, the numbers out of the United Kingdom are high: a recent report indicated that of those who drank before they conceived, 55 percent drank when they were pregnant. Those most likely to drink included women aged thirty-five and older (61 percent), compared with those under twenty (47 percent).

Last year, the U.S. Centers for Disease Control and Prevention (CDC) reported that one in thirteen American pregnant women said that they drank, with the highest prevalence for those aged 35 to 44 (14 percent), white (8 percent), college-educated (10 percent), or employed (9 percent). Pregnant women who were employed were nearly 2.5 times more likely to engage in binge drinking than their unemployed counterparts. Women who were bingeing reported doing so an average of three times a month, consuming six drinks per episode.

Meanwhile, a 2012 study in Australia reported that 47 percent of women consumed alcohol while pregnant, before their pregnancy was confirmed; and 19.5 percent consumed alcohol while pregnant, knowing they were so. Older women with a higher household income were more likely to drink after they learned they were pregnant. Ninety percent of drinkers under 25 quit drinking once their pregnancy was confirmed, whereas only half of those aged 36 or older did the same.

Says American scientist Phillip May, arguably the leading researcher on FASD prevalence: "How do you get over the middle- and upper-class arrogance around FASD? 'I can have three or four glasses of wine a week, and my child will be fine!' 'These are lower-class problems. They are the ones with public health problems.' And they likely *will* be fine, too. Their children will have an IQ of one hundred. They may have a menial labor job—when they could have been an engineer."

Currently appointed as a research professor at the Gillings School of Global Public Health at the University of North Carolina at Chapel Hill, May says that both FAS and FASD are more prevalent than previously estimated: "Roughly two to seven children per thousand have fetal alcohol syndrome in North America and Europe, and two to five percent have FASD. In the United States, Canada, and especially Europe, between fifteen and thirty percent of all kids have been exposed to a significant amount of alcohol as fetuses. They may be born normal, or close to normal—but who's to say if they would have been geniuses had they not been exposed? The moderately exposed kids might have adequate nutrition, stimulation, and education—so for many aspects of life they may perform in the normal range. It may show up in poor judgment, an inability to do math, or more complex tasks of reasoning. Meanwhile, in Scandinavia and the United Kingdom, up to twenty percent of mothers are drinking at very high levels in the first trimester. And many are drinking at levels higher than they report. I can tell you: fetal alcohol spectrum disorder and fetal alcohol syndrome are far more common than I ever dreamed they would be."

If you look at special populations, the numbers of those with FASD are high. In the United States, up to 35 percent of adoptees have been reported to have FASD—the majority being from Eastern Europe. In Sweden, up to 30 percent of adoptees from Eastern Europe are reported to have FAS. The proportion of those with FASD in correctional systems is also very high: up to 23 percent in Canada.

One of the keys to this story is delayed pregnancy recognition. You can ask a woman if she drank during a particular pregnancy and she will say: "I quit as soon as I had a positive test." Much can happen before that pregnancy confirmation. According to May, many children may not reach their full potential: "A lot can happen in the eighth, tenth, twelfth week. That exposure seems to lead to cognitive and behavioral problems. There are all sorts of epigenetic reactions that we don't understand. If a man drinks heavily, he can have abnormal sperm. A kid can get a deficit from both sides."

Personally, I'm fond of Jody Allen Crowe's initiative: installing pregnancy test dispensers in bars. Last summer, the founder of Healthy Brains for Children placed a dispenser in the women's washroom at Pub 500 in Mankato, a college town in southern Minnesota.

For three dollars, a woman could find out if the drink she was about to order was a good idea or not. Smart innovation.

So too is the labeling of liquor bottles—but May is skeptical. "Bottle labels have a minimal effect—for light drinkers. If women are already drinking, it means nothing. But the next generation? They will grow up *knowing* that alcohol and pregnancy don't mix. It's just like seat belts—seat belts save lives. We accept this now. They will know that you should stop drinking before you try to conceive."

According to Popova, more than 50 percent of pregnancies are unplanned in the developed world, and closer to 70 or 80 percent in some countries in the developing world. She says that five years ago, we believed that drinking during the first and last trimesters of pregnancy was the most harmful. The most recent evidence from animal studies demonstrates that the fetus is vulnerable to alcohol at all stages. She adds that women should abstain from drinking during breastfeeding: "Alcohol consumed by the mother passes easily into her breast milk at concentrations similar to that in her bloodstream. Therefore, alcohol goes directly to the baby, which might result in impaired mental and motor development, changes in sleep patterns, and growth deficits. FASD is largely preventable. What woman wants to damage her child? Unfortunately, the public tends to stick with ideas that they like—or are convenient."

As part of the international study guided by the World Health Organization on the global prevalence of FASD, Popova will be leading a school study of eight thousand children aged 7 to 9 in the Greater Toronto area of Canada. The research team will identify those children who have growth deficits, learning disabilities, and behavioral problems. These children will then have a dysmorphology assessment, measuring their facial features. They will also undergo psychological and developmental testing. Their biological mothers will be asked about nutrition, stress, alcohol, and tobacco use during their pregnancy with their children in the study. If a mother does not confirm she consumed alcohol during her pregnancy, those assessing her child will not be able to assign an FASD diagnosis, no matter how obvious the case. The hurdles are immense.

Sarah Mattson, who is associate director of the Center for Behavioral Teratology at San Diego State University, focuses on better ways to determine whether certain children were exposed to alcohol as fetuses. She and her team look at adopted children or children in foster care, ones for whom there is a social worker's report or some other confirmation that the birth mother used alcohol while pregnant. "We do a slew of neuropsychological testing to find out what the child's deficits are. We are looking for a marker that says if the child has x, y, and z, he has a high chance of having been exposed to alcohol. With every study, we're getting a little bit closer to finding one." I ask Mattson: does she believe there are children in the school system who are labeled learning disabled, who actually have FASD? "I am *sure* there are kids who are not being recognized," she says. "Absolutely. Having the facial features is not important—the exposure to alcohol is, and whether they show similar behavioral and cognitive features."

"I binge drank through my pregnancy," says Janet Christie, matter-of-factly. "I really loved drinking. I knew when I was pregnant that it wasn't good to drink. I was so ashamed. But I had no one to talk to about it."

Sitting in a sunlit corner of Vancouver's Westin Bayshore hotel, Christie has agreed to talk about being the birth mother of a child with fetal alcohol spectrum disorder, on one condition: that she can continue folding brochures throughout the interview, small pamphlets advertising her services in training addiction recovery coaches. With that settled, she launches into the full story of how her drinking started, and how she got sober twenty-five years ago.

"When I was really young—only eight—I tried to drown myself. I was shy, timid, internally isolated. Then, when I was thirteen, I had my first drink. I will never forget it: before that, I felt worthless, so empty. A hole where my heart should have been. That first drink was amazing—I loved it. It didn't matter that I woke up in a stranger's car with a sore crotch and blood on my underwear—I was sexually assaulted."

Fast-forward to her sobriety. "I was one week sober, I'd found a recovery support group, and the phone rang. It was the police. They had caught my son, who was twelve, in a crack shack. I didn't even think he played with matches! This was my introduction to recovery. In those horrific years, I was so afraid he was going to kill somebody—I felt like I was in the front line of a war.

"I had known since he failed grade two that I might have caused his problem. And that's when things started to go awry. He had undiagnosed FASD. I read the research papers, and I told the principal, 'I think I caused this.' He said, 'Just go home and forget about it'—as nonchalantly as 'There's a toilet roll that needs to be changed.' Then he said, 'I bet ninety-five percent of the kids here have FASD.' I was giving talks at treatment centers, telling my story, and a professional said to me, 'You caused it—deal with it!' But another counselor said, 'Tell your son he doesn't have to live like this anymore. Get him diagnosed.' So, when he was fifteen, I did. He cried."

At the time, her son couldn't read. "He would skip lines and not realize it," she says. He couldn't add, and he couldn't comprehend what he was being taught at school. He was ostracized. Without identification of what he had, he had no support. "He felt put down by teachers, blamed for not trying hard enough. He wasn't connecting any actions with consequences. He was full of rage. I was worried I would open the paper and learn he had killed someone."

Christie, who lives in an ocean-view house in Sooke, British Columbia, speaks publicly about her life in an effort to help other women stop drinking while pregnant—and to try to reduce the stigma around the subject of addiction. "The world doesn't always feel like a safe place when you're a birth mother of a child with FASD," says Christie. "This may be the most stigmatized area of a very stigmatized subject. It's hard coming out—like you're a leper. No wonder we don't have more women talking about it. Years of tears—how many times have I told my story and cried through it? But there is a freedom in this."

Polished and well spoken, she knows that she doesn't fit the stereotype of an FASD mom—and this is part of why she's taken on the role of speaking out. "We need to break the silence barrier. Because it's alcohol and it's a revered substance, it doesn't get talked about in our society. The myth that this only happens to certain women is wrong. It pushes middle-class women even further into the closet. White women just pretend their kids have learning disabilities."

For years, Christie's son skipped class, and was kicked out of several schools. Christie joined a parent group for those with FASD children. Her son pawned her jewelry. He had drug debts. He would disappear for days and then end up with some criminal. He would constantly break his hands, punching walls, breaking doors, coffee tables. She kicked him out. "I tried to do the tough-love thing," she says. For a while, he came back home and lived in her garage, sleeping in her car and cooking on camping equipment.

Over time, with her support, he got on the right track. Christie sees his life as a success, although he struggles. Today he is no longer using drugs, is employed in construction, and has a stable, loving relationship with his girlfriend. His twelve-year-old daughter, whom he sees regularly, is being raised by her grandmother on her mother's side. "We have to celebrate the uniqueness of brain differences, and my son is a perfect example," says Christie. "He lives a very functional life, pays his taxes. Many people, if they are properly supported, do fine. We have to measure success individually. I am so inspired by his tenacity. He knows why he struggles. But he has a great sense of humor and a great outlook on life. And things are gradually turning around for him."

Christie's passion is a program she launched in 2004. Called Moms Mentoring Moms, it is a support group for women struggling with addiction while pregnant; some have lost custody of their children. "Peer support is really fundamental for anyone wanting to overcome an addiction," says Christie, and peer support is what the program offered for women who wanted to stop drinking—support without judgment. Launched with eighty thousand dollars from a British Columbia nonprofit agency and the provincial Ministry of Children and Family Development, the group provided a weekly drop-in for women, as well as a mentor to accompany them to any appointments: navigating the search for housing, dealing with social workers, applying for welfare, visiting the food bank. Many women found sobriety through the group. The program ended when the funding ran out after a year.

What Christie was doing had Nancy Poole's full support. "The main issue around the mothering part is the tremendous shame, stigma, and fear that their children will be taken from them if they say that they are drinking," says Poole. "It's almost overwhelming for most women to think about stopping drinking. There are lots of things we can do to be terribly supportive: buying vitamins, getting strollers, offering multilevel support that is really valuable. We need to offer multidimensional help, to get at the things that are behind their use. Mentoring programs are so important. There is so much potential for us to do better: helping service providers interact with women in a respectful way."

Today Christie trains volunteers to do just this, and she is trying to raise funds to relaunch the program. In addition, she uses her own website to raise awareness of the dangers of drinking while pregnant. "Recently, studies have appeared suggesting that women need not worry about consuming low levels of alcohol during pregnancy," she writes. "I find this very disturbing and question why, as a society, we are spending money trying to prove it is all right to mess around with an unborn child's human potential."

Most of all, Christie battles society's notion that "addiction is a moral rather than a sociomedical issue." Personally, she says, "I feel blessed to have made it out of that big black hole of addiction. I feel this is my calling, to do this work, and it gives me a great deal of pleasure to see how a little bit of effort can make a big difference in another mother's life. I have a passion for the moms. In the end, shaming and blaming comes from a place of misunderstanding. It's a useless waste of energy. It's so much easier to point a finger than hold out a hand."

Holding out a hand is exactly what Margaret Leslie and her team do at Breaking the Cycle, primarily a children's center in downtown Toronto: a one-stop-shopping service for pregnant and parenting mothers of children six and under, all with substance use issues. Every Thursday, Dr. Gideon Koren, a pediatric toxicologist at the Hospital for Sick Children, comes to the center to offer an FASD diagnosis clinic. "It's a very unique setting because it's an opportunity to see the birth mothers," says director Leslie. "This is a population of women who aren't normally seen. This is a safe place where they can ask all the questions they haven't wanted to ask another physician. These mothers want their children to have success in school in a way they never did. Every single mom's biggest worry is the effect of the prenatal exposure."

A woman cannot be intoxicated or high when using the center. Breaking the Cycle offers addiction counseling, and it also has a partnership with Toronto Western Hospital, which has trauma services, plus mental health and addiction services. "We are an attachment-based program," says Leslie, waving to a vibrant young mother arriving with her newborn son. The woman is here for parenting class. "Our ultimate goal is to promote attachment. In the end, the stronger the attachment, the higher the protective factor in mothering."

12.

The Daughters' Stories

GROWING UP WITH AN ALCOHOLIC MOTHER

Man hands on misery to man.
It deepens like a coastal shelf.
—PHILIP LARKIN

Serious drinkers are like serious eaters or serious
noneaters. They are like serious drug-addicts. Their addiction
holds a spell over them which acts as some powerful
secret at the center of everything they do.
—MARION WOODMAN

It has been said that families are like mobiles: when one person shifts, the others will as well. When one person drinks, each member of the family is thrown off balance. If the drinker drinks, all suffer; if she gets sober, the entire family feels the impact. Both require enormous adjustment.

When my mother returned from a stint in rehab, our whole family walked on eggshells. We were accustomed to a mother who slept much of the day, paced the halls at night, ice cubes tinkling. We were used to her 2 a.m. rants, her anger, her quixotic temper and cutting sarcasm. We were not used to this smaller, shaky person, who looked painfully vulnerable, remorseful, and soft. She was like a newly hatched chick, feminine and lovely—and utterly foreign to us. It was like having a houseguest: she moved around her own kitchen with tentative steps, articulating what was in her heart. With my whole being, I wanted to help her. I felt as if E.T. had just arrived: I liked this new foreign being, but I was not sure how long she would stay.

Turns out, not too long. My mother had trouble with the sharing in her follow-up groups. "I am *not* going to talk about my family with strangers—I wasn't raised that way," she'd say when the group coordinator called to remind her of an upcoming meeting. Before long, she stopped attending the after-care portion of her program. In a very short time, our other mother reappeared: nasty and sad and broken-hearted with how life had shortchanged her. Sleep was disrupted, dinners were difficult. This was the old normal.

Still, I never looked at her the same way again. Once I had been reintroduced to my softer mother, the mother of my early childhood, I never gave up hope that she might reappear.

It took a long time: decades, to be honest.

My mother used to drink and dial: it was her favorite pastime. Once, in the early days of my magazine career at *Maclean's*, she phoned the editor of the magazine—Peter C. Newman—in the middle of the afternoon. I don't quite remember how he shared this news with me: the fact that she had reached our editor in chief was mortifying. On another occasion, she got through to the art department, and invited one of the designers to our summer cottage on Georgian Bay. He thought it was sweet, but random. She is very charming.

As a teenager, I never grew accustomed to her calling the parents of friends or a new boyfriend, often past midnight. But by the time she started calling the police, I was in my twenties, and I found it both irritating and vaguely amusing. More than once, I had a date interrupted by an officer explaining to me that I hadn't called home in several weeks. They always looked skeptical when I said not to worry: my mother was exceptionally persuasive.

The worst, by far, was a wedding shower that I gave, fresh out of university, in our home. An hour before guests were to arrive, my mother disappeared. My sister and brother and I knew she had been drinking, heavily, for days. No one could find her. Halfway through the event, she appeared in a long hostess gown, poured herself a drink, and held court for the rest of the afternoon—until the alcohol hit her and she told "all the little bitches" to leave the house immediately. When it came time for my own wedding, her behavior was no different—except she had a black eye from falling: she applied plenty of makeup and posed for pictures. That day, I was more focused on her presence than that of my new husband, which was a shame: I was twenty-three and still enmeshed in my family drama. It would be years before I unhooked myself from regular upset, through therapy and distance.

All through my twenties, there were trips to the hospital, violent episodes, cuts and stitches: alcohol-fueled fights or accidents that broke your heart. My brother, John, took the brunt of it. As the youngest, he was at home alone for several years, once my sister, Cate, and I headed off to university. But we all suffered emotionally. How could you not? We spent years trying to hide the truth from our friends. Once we began to marry, others were exposed to the unvarnished reality. "My mother has something seriously wrong with her," I said to my future husband. "Probably an African throat disease."

"You know that's not true, don't you?" he said gently, firmly. "She's an alcoholic."

I still remember the shock of his saying it out loud, of articulating what we were forbidden to reference: in our household the proverbial elephant was alive and well in the living room, in the bedrooms, and just about everywhere else. Only my brother was subversive enough to make fun of my mother's behavior, often mimicking her struggling to get through doorways, like a conductor on a runaway train.

His antics helped relieve the tension for my new husband. This helped at the dinner table. But there was no preparing a newcomer for an overnight visit at our cottage. I said very little to Will, but I knew it would be difficult. At night, my mother had two modes: the ranting mode and the musical mode. Ranting was angry; musical was maudlin. If it was musical, we would be treated to an endless evening of *Chariots of Fire*—or, as my brother called it, *Chariots of Firewater*. If it was a ranting episode, she would deliver a soliloquy on our faults.

On Will's first night at the cottage, it was a ranting evening. When my mother's performance began in the upstairs hall, he whispered: "She does this every night?" I nodded. "You're kidding." "Just roll over. Whatever you do, don't go out there."

He tossed a towel around his waist and headed into the hall, saying in his best no-nonsense voice: "Maxie, others are trying to sleep. Go to bed." And she did, just like that. In four cottage bedrooms, each of us silently cheered. Of course, it only worked once. The next night, she let him have it.

Years later, when Jake arrived on the scene, she had put hard liquor behind her. Still, Jake summed up the scene perfectly. He called her a lion-tamer: "She cracks the whip, and you all jump." He was right. She was a lion-tamer, and we were her four cubs, my dad included. We all played our roles: my father was silent, I fought back, my sister hid, and my kid brother joked. Codependent to the hilt. We had it nailed.

Today, with my father gone, my softer mother only drinks in the evening. She limits her intake to two Diet Coke spritzers, pouring both at 6:45, depositing one in the fridge. At 7:00 she sits down to watch *Wheel of Fortune*. At 7:30 she watches *Jeopardy!* If we're together for the night, we'll start a game of Scrabble. By 9:00 she'll serve dinner. By 10:00 she's in bed. I could set my clock by it: it's endlessly reassuring, this new regimen, and I never get tired of seeing her in the upstairs hall of her lovely home, in her pretty flannel nightie. "Good night, Mum, sleep well." "Do you find the house a little cold?" "No, I think we're fine, Mum." "Well, sleep well, dearie. There's an extra blanket if you need it. It's so nice to know you're here." "I love you, Mum." "I love you, too, darling." Silence. Endless, comforting silence, until the sun comes up.

Still, I can't help but bristle when the wine comes out. We all do. Even though it has been several years since there was any serious trouble, I am still adjusting to this new reality. Whenever she puts that drink to her lips, I can't help but wonder: will this be the one that snatches her back, this mother I have grown to trust? It never does.

Most tragically, just as my mother adjusted her drinking regimen, my father slipped into alcoholism. In retirement, he was a secret drinker—or not-so-secret: every member of the family knew that his trips to the garage, to "fetch something from the boot," meant opening the car trunk to take a long swig of scotch. The trunk was his bar: this never changed. "For God's sake," my mother would say, "just use a glass, John." He rarely did. He was his own man, and he remained that way until he died: loving my mother, loving his mining research and his stocks. Eventually he switched to vodka, as most problem drinkers do, taking a bottle with him as he walked their dog on a summer afternoon, returning loaded to the gills. No one knew what to do: here was our so-called sober parent, following the same addictive path. It ended up killing him—which left all of us perplexed and inconsolably sad. He was a fine man with a brilliant brain, a wealth of stories stocked up from years of international adventures. He never lost his tender heart, and his absence is a presence we have adjusted to, with difficulty. My mother has borne this with grace and exquisite dignity.

For me, my father's absence and Jake's are twinned: they disappeared within months of each other. That double loss has freed time for my mother and me to travel together. In summer, we play Scrabble for long hours by Georgian Bay, our dogs sleeping at our feet. Once in a while, she will mention her drinking days and say how sorry she feels about "what I put my family through." I'll pass her the bag of tiles and tell her how happy I am that we get to spend time together. "I think I might pour myself a drink. What time is it?" "Six thirty, Mum. Go ahead?" "What will you have, dear?" "Nothing, Mum, I'm happy with water." "I'm so proud of you, darling. How many years has it been?"

This is how it always goes. When she returns with her spritzer, we continue our game. "Don't you hate to eat alone?" she will always say. "Yes, Mum, I do." "I always knew you must be lonely, living alone, but I never really understood it until I had to go through it myself." She'll light another cigarette, and contemplate her letters. "I have a great word, but I'm not sure how long I should hang on to it." "Hang on," I'll tell her. "I'd like to," she'll say, with her girlish smile. "You might as well, Mum. You just never know."

"How have you been able to forgive her, after all she put you through?" This is the voice of my friend Barbie. Barbie was around in high school, when things turned bad. On many weekends, I escaped to her family's large home overlooking a park. "It can happen to the best of us," I say. "But your drinking and hers were not the same." "She knows that," I say. "Besides, I love her." "You're a good daughter," says Barbie. We leave it at that.

We're all good daughters: that's the summation of Robert Ackerman, author of the first book on children of alcoholics in the United States, as well as *Perfect Daughters: Adult Daughters of Alcoholics.* We're comparatively rare: it's not that common to have an alcoholic mother. Of those raised by alcoholics, 60 percent have an alcoholic father; only 20 percent have an alcoholic mother, and another 20 percent have two alcoholic parents. According to Ackerman, daughters raised by alcoholic mothers have different issues than those with alcoholic fathers. Since they're in the minority, daughters of female alcoholics rarely have a friend in the same situation. As a result, they tend to feel more isolated. Most grow up feeling anger, disgust, and disappointment. A smaller proportion take a protective role with their mothers, denying their parent is alcoholic. And most remarkably, they all tend to accept that their mother is alcoholic much later than those who have an alcoholic dad: close to nineteen years old versus thirteen.

I was one of those daughters. I was in my early twenties before I could fully accept that my mother wasn't suffering from an African throat disease or cancer. Peggy McGillicuddy, who grew up in California, was no different. She was thirteen when her mother went to treatment. There a counselor told her that her mother was an alcoholic, and she got really angry. "I was like: 'We're done here!' I spent the next twelve years denying that my mother was an alcoholic."

Still, she had lived with the reality. "I didn't realize it wasn't normal for a mother to sleep a lot in the day," says McGillicuddy. "I remember being incredibly careful as a kid about who came to my house because I wasn't certain what it would be like. I knew there was something incredibly bad about my mother. The stigma of having a mom who was an alcoholic was profound, damaging to me."

Like me, McGillicuddy was a classic case: she had straight A's in school. "I was the perfect kid," she says. "But after college, I fell apart. I had depression and an eating disorder—and I was hospitalized. Not once did a therapist connect my depression with what I went through as a child."

Says Ackerman, "Of course, it's not the drinking that drives people crazy. It's the unwanted behavior, the dysfunction." Is there an upside to all McGillicuddy went through? I know the answer before she opens her mouth. "Yes," she says. "I can walk into a room and figure out *immediately* what someone needs from me emotionally." I nod my head: our antennae are perfectly attuned for trouble. It's a gift for life—if you want to look at it that way.

Closing time, a British pub in Chelsea. Three people are standing at the bar, two tall men and an attractive dark-haired woman, all in their late twenties. The security guard approaches them. "Madam, sirs, it's time to leave." The young woman gives him a little "cheers" salute, and raises her wine—except she misses her mouth. "Oh no! I slopped on your socks!" she says, stumbling.

"Madam, it's time to leave," says the guard.

"Oh, come on, it's only Thursday! Tomorrow's Friday. Then it's the weekend!"

"Madam, it's Sunday night, and we're closing."

The young woman guzzles what's left of her wine. "Can I take this home?" she asks, lifting the bottle. "No, madam, I need to lock the door." "Let me just finish this. You know, I think I'm tipsy!" She stumbles, and one of the tall men catches her, escorting her out.

My guest and I watch this scene, putting on our coats as we too leave Phene, a smart gastro pub with a welcoming neighborhood feel and a posh menu. Both daughters of alcoholic women, we look at each other, uncomfortably. Nothing we haven't seen before, but it makes our skin crawl.

Caroline, as she wants to be known, has chosen this restaurant in which to tell me her story. Like McGillicuddy, Caroline is a high achiever. Unlike her, she had no illusions about her mother's alcoholism. Slim and poised, with a winning smile, she lost her mother to alcohol a little more than a year ago, and the pain is very fresh. Still, at thirty, she is able to find her equanimity as her story unfolds. "My mother didn't want to be here. Even though she had all of the right things, she was desperate for an exit. Alcohol was an escape, and she got what she wanted. It's terrible to say, but it was a relief for her to go. The closest people could not get through to her. I understand how all this came to be so catastrophic—but I will probably never fully understand it."

Caroline's mother died a year ago, at fifty-eight. An accomplished office manager, she had been the breadwinner of her family and a devoted mother. She worked hard, and made good money. However, when her own parents died, she fell into a depression and was reluctant to see a doctor. She began to drink heavily, and sneakily, just as her own father had. At first it was wine, but as the years progressed she turned to spirits, drinking mostly in secret. Socially, she would stick to the script. At home? She would drink anything she could, hiding bottles around the house.

Caroline remembers being a child at relatives' homes and having her father insist: "We have to go now!" She would protest: "Why? We're having so much fun!" "*Now*," he would say. "We have to leave *now*." Home they would go, and Caroline's mother would begin her second shift of drinking, beyond the eyes of extended family. "My father preserved her dignity, and our dignity as a family," says Caroline. "It was important to him."

When she headed off to university her mother's drinking got worse. On top of the alcoholism came anorexia: Caroline's mother shrank from two hundred pounds down to ninety. Skeletally thin, she still had a drinker's belly. Food disgusted her. She spent her free time watching celebrity shows, and avoiding eating. Meanwhile, she began to fall a lot, once down an entire flight of stairs.

Then came what Caroline calls the "tipping point": her mother lost her job of twenty-three years. "I'm sure she was drinking on the job." She became a cruise ship consultant—but her salary was small compared with what she had earned before. She felt like she was "nothing." "She began to feel guilty about her drinking," says Caroline, "and the only way to get rid of the feeling was to have another glass."

When Caroline returned home for Christmas in 2006—from England, where she was doing her master's—her parents came to meet her at the airport, but her mother couldn't stay upright. "That Christmas was very difficult," says Caroline. "My mom had dementia. She threw food across the room."

On another occasion, her mother fell and had a seizure in Caroline's arms. Caroline and her father called 911. By the time the paramedics arrived, her mother was lucid. "I'm fine," she said. She was coherent. "The paramedics wouldn't take her," says Caroline. "My father and I were both weeping, begging them to take her. They would not. We discussed getting a court order to force her into rehab—but my mom could be remarkably coherent. It never happened. We were terrified."

Things grew worse—but still Caroline's mother would not get help. She became more difficult. "There were extreme, scary times," says Caroline. "Alcohol totally transforms people. She isolated herself completely, pushing all her friends away. She just self-destructed."

"When people ask, 'What did your mother die from?' I have trouble saying, 'She was an alcoholic,'" says Caroline. "I don't want to say, 'She was anorexic, she had liver cirrhosis and extreme dementia. She drank herself to death.'" Caroline can say none of this without weeping. Much of her life is spent dealing with her grief. "It's like a tattoo: it never goes away. I think about her every day. I want to stamp my foot and say, 'This is not the way life is supposed to be.'"

Says Ackerman, "One of the top five things that adult daughters of female alcoholics feel is anger that they were not provided with a role model. 'You never showed me! Do you want to explain to me how to be a mother—I've never seen it done.' Parenting is a big issue, and the biggest issue is relationships."

Caroline is no exception. "My mom didn't teach me a way to be a woman. I have to look to other people—and it's a huge struggle to teach yourself things that you should have learned at home." Like? "Self-care."

This answer pierces me to the core: I live this truth. Self-care is a learned behavior, when you have an alcoholic mother. Kate, a vibrant owner of a small Manhattan investment firm, knows this truth as well. She works out with a trainer two times a week, travels regularly, makes time for friends. She plays squash, tennis, golf. "I refuse to be like my mom—crippled, a depressive," says Kate.

The eldest daughter of two lawyers, Kate has two younger brothers, both of whom she tried to protect from the reality of a stay-at-home drinking mom. "My dad would say, in retrospect, she drank from the beginning of their marriage." To keep the peace, Kate became a classic "perfect daughter": "I was doing a lot of chores from an early age, and I would find vodka and gin bottles in the closet." She also became a good cook. "It was either that or starve."

Kate is aware that her mother had a nervous breakdown in her teens and was given shock treatment "in a pretty brutal way. She had already had a bad run-in with mental health when she married. She was a depressant, and she was self-medicating. I would come home from school and she would be in her bed, asleep. The more she drank, the harder my father worked. There were constant battles, although he stuck by her. He loved her, unconditionally.

"Christmas was a nightmare: once she tried to cook a frozen turkey. Ultimately, she had a car accident and they took away her driving privileges. In the end, she started drinking vanilla. My brothers would say, 'You'd think we were in the bakery business.'"

To this day, Kate is troubled by the fact that she has never married. "I suspect that all this history is why I am alone," she says wistfully. More than once in our interview, she returns to this subject. Clearly, this haunts her. It haunts Caroline. There is a feeling of being cursed that comes from having an alcoholic mother: a curse that one can never quite shake off. It's as if someone forgot to give you the essential manual to life. This feeling is common, and it never fades.

Have a parent who drinks, and you are more than four times as likely to be concerned about your own drinking or drug-taking, says Ackerman. This fact is well known to Dr. Vera Tarman. As medical director for Renascent treatment center in Toronto, Tarman is well schooled in others' stories of alcohol and addiction. But my interest, on a cold January Sunday, is in her own story. I arrive in her cozy book-lined study to find an intellectual woman in her fifties, small dog at her side; she is ready to share her entire story.

As she sees it, Tarman's relationship with alcohol began when she was seven or eight. She remembers her mother, a psychiatric nurse, falling, getting groggy, acting strangely. Early on, her mother was fired when she was caught stealing pills on the job. This began a series of new jobs, more firings. By the time Tarman was ten, her mother was in bed much of the time, drinking daily, smelling of alcohol. "She was not a functioning alcoholic," says Tarman. "The progression was very quick."

When her parents separated—"my father had had enough"— Tarman's mother had an affair with another woman. Although Tarman was "aghast," this relationship had one major benefit: the woman introduced the young girl to books. But soon, as with Tarman's father, the woman had had enough: she too left. This prompted her mother to make the first of many suicide attempts—"very distressing to live with," says Tarman. Financially destitute, the two then moved back with Tarman's father, an arrangement that pleased no one. "It was my job to deal with her alcoholism," says Tarman, "and lots happened. Many visits from ambulances, firemen, neighbors. She would scream when she was drunk, there were visits to psychiatric hospitals. All the time, I would be worried: 'Where is she? Am I safe? Has she started a fire, smoking in bed?' If I didn't find my mother's bottles, my father blamed me. Her drink was Five Star whiskey, and I collected the labels, like badges—symbols of my mother. I could not throw them away."

When Tarman was fifteen, her mother's suicide attempts increased. Finally, she was successful, escaping from a psychiatric hospital and drowning herself in a pond, her bottles nearby. "I was horrified because I was still very close to my mother," says Tarman. "She was constantly apologizing to me: 'I will stop tomorrow.' I did not hate her. And when she died, I lost my closest human contact. There was part of me that said, 'Don't leave me behind! I want to come with you.' But when she died, I was also greatly relieved because it was the end of a long hell.

"People often say that when a person goes through a hard time, they only need one other person—and that person was Inge, a friend of my mother's. She was my savior. 'You can be anything you want,' she would say." She gave Tarman hope.

Tarman blamed her father for her mother's death—he had had an affair, also drank heavily, and was frequently charged with drunk driving. When she was seventeen, she had had enough: she moved into subsidized housing and began to experiment with drugs. At eighteen she went into treatment for nine months, emerged to finish high school, and headed to university, where her drinking escalated. She also developed severe bulimia. On graduation, she traveled in India and Europe, where she became "very depressed and suicidal. Sylvia

Plath was my hero, and I knew that drinking was getting me there. It was very seductive."

At medical school, Tarman cut back on her drinking. But once she had graduated, she began to appreciate fine wines and better drinks. Over the years, her drinking escalated, as did her food issues. Eventually, she ballooned to 240 pounds. "I would throw up until my mouth was bleeding," says Tarman. Over time, she switched into addictions medicine and addressed her eating, adopting a low-fat diet, losing one hundred pounds. But the alcohol remained an issue—even though she was "aware of the hypocrisy of telling patients not to drink and then drinking myself." Between 10 p.m. and 2 a.m.—once her partner had gone to bed—she could let go, drinking several double martinis, her favorite.

Eventually, she eliminated alcohol, too, quitting on her own. That lasted four years but it brought her no happiness. "It worked," said Tarman, "but it wasn't a happy place to be. Four years in, I thought: 'I can't live like this. I work all day. I can't have a nice martini or half a cake.' I turned fifty and got to a very dark place again. I became very reckless. I thought: 'Fuck it. Is this all there is?' Before, what had pulled me out was my career: I was very ambitious. But despite the fact I had a successful work life and a happy home life, I could not find my way out of depression. I was in a classic crisis: I had achieved my goals and I expected big fireworks in life. There were no fireworks. I was feeling quite starved of little pleasures. It was just an exercise in frustration—so I decided to go for obliteration. If I died, so what? The romance of this kind of death, like my mother's, was very seductive."

Tarman drank for six months. "I was more reckless than I had been in the past. Within six months, I had to make a decision: whether to stop or to throw everything away. A couple of glasses of wine wasn't enough. The little sweet spot would last only five minutes, and I wanted to be totally wasted. My suicidality got more so. I added drugs."

Tarman hit the wall, calling her medical support help line. Eventually she found a peer support group. "There I felt melt-in-my-seat safety, even before people began to talk. Nothing cuts it like this group—not therapy, not meditation, not church. I am more needy than what the universe can give, and this group soothes that need. Reality doesn't bother me as much as it used to. It's as if I have been waiting for my mother to come back, and she never has. This group is the only place that soothes the ache. This group is my daily bread—and 'You are no longer alone' was the first slogan that comforted me."

I wondered: is she triggered in her work, dealing all day as she does with addicts? "Yes. It is always there when a patient says how drunk they got. I still really miss those first two martinis. I feel a beckoning finger. Although the volume of the voice changes, I am constantly battling with 'I want to have a blast. Reality just doesn't cut it!' But if I had another drink, it would open a door I am not sure I could close again.

"I see myself as a survivor," says Tarman. "Yes, I have vulnerabilities. But I have strengths. I read people better than most. I respond well in a crisis. I have been with my partner, Cathy, for twenty-five years.

"Still, my mother dominates in so many ways—I am always aware of that legacy. I dedicated my book to her, instead of Cathy, my partner, whom I love and who deserves that recognition. My mother was a geriatric nurse, and I chose medicine. The specter of my mother's death continues to dominate as a regular living ghost." Even though she has found contentment through her recovery work, there is an invitation to "come." Says Tarman, "The familiarity of her death and the fact that she died before I reached adolescence—and was old enough to hate how our life together affected me—make it an open invitation. 'Come, follow me!' There's a wonderful memoir by Julia Child where death is perceived as just getting off the bus. This is definitely a residual daughter thing. I just think: 'when it is bad enough, I will just get off the bus, come back home.'"

Weeks after our interview, I received an email from Tarman: she is concerned that she had presented her life as too sad. It's a thoughtful note, and I consider what she is saying. She's right: it is confoundingly sad to deal with an alcoholic mother when you are young. And I can't imagine losing a mother to alcoholism, to death, forever, as she has.

How to explain the unshakable sense of bonding with an alcoholic mother? You hear it in her story, and I feel it in mine. My mother and I are twinned: so yoked that I can't imagine life without her. I remember nursing her on a hot summer weekend, when she had taken Antabuse—the pill-form deterrent to drinking—and then poured herself a rum and Coke, and another. Shaking with nausea, clammy-skinned and gray, grateful for the company and the help, she sat on the couch and thanked me for leaving the rest of the family at the cottage, and driving to be with her. She was too sick for Scrabble. We just talked. Yes, there were those moments—plenty of them.

I often feel it is my job to live my mother's dreams, to travel in her footsteps and then forge ahead where she was restrained, to report back to her from the world beyond.

I love sharing our journeys. And I love holding her soft hand in mine as we cross the rocky shoreline of Georgian Bay each summer, toward the smoother footing of the deeper water. Together, we are survivors. We both know this. The pact is silent and profound. We have been through the wars, together.

PART THREE

Healing

13.

In Which Everything Changes

GETTING SOBER, STAYING SOBER

*If you can kill the right thing—the old way of adaptation—
and not injure yourself, a new energy-filled era will begin.*
—ROBERT A. JOHNSON

Massachusetts, Winter of 2008

The first Friday evening at rehab, the parking lot empties before
five. The director has a slightly guilty look as he puts on his coat. Or
at least, I imagine he does. The BMWs and the Lexus and the white
Jeep Cherokee all pull away, one after another, heading in different
directions, back to the bedroom communities and families and
weekend plans.

Now, what am I to do? I am forlorn. Two whole days without the
rest of the crew, without the good doctor who has just told me I
likely have severe post-traumatic stress disorder—PTSD. I feel drop-
kicked into oblivion.

"I'm getting out of here," says Claire, shoving a man's woolen hat
on her head. "These weekend people don't know anything." Claire
is here because of a DUI. Unlike the rest of us, she's here on a court-
ordered stay.

"Don't worry," whispers Jane. "She'll calm down with a little aromatherapy." I'm confused. "Smoking. She's gone for a cigarette."

"Go fuck yourselves," says Claire, slamming the door.

Fifteen minutes later, she's back, brandishing a vodka bottle. "Look what I found by the side of the road," she says, grinning. "I saw a ding in the guardrail, and I knew this baby couldn't be too far away!" A half inch of clear liquid is sloshing back and forth in the bottle.

"Put it down, Claire," says Jane.

"I'm not going to drink it," says Claire. "And Jane? Go fuck yourself."

That night, heading to an AA meeting, I squeeze into the back row of the so-called Druggy Buggy, making sure I'm as far away from Claire as possible. Marilyn is at the wheel. She has big hair, big makeup, and a knitting bag, with a half-completed baby blanket, pale green and yellow and pink. (I feel sorry for the baby.)

Marilyn is fresh from Las Vegas, where things, apparently, went very poorly. This I know because one of the night staff shared it the previous evening when I couldn't sleep.

Marilyn whizzes us down the mountain, past the night skiers, past all the little glowing houses. I try not to listen as Jonathan discusses the relative differences of breast implants on Caucasian and Asian women. "They look ridiculous on Asian women. The scars are too obvious." Jonathan is in his early twenties.

"Shut up, Jonathan," says Claire. "None of us care."

Jonathan looks confused. I know his problems were cocaine and vodka. Maybe more. I try to focus, on the stars that most houses have nailed beside their front door. Maybe it's a New England thing. It looks corny, like some grade school teacher went a little berserk.

I try to focus on the meeting ahead, where—I have been warned—an old man will be pacing up and down the aisles, a knife swinging from his belt, talking to himself. "Big swinging dick," says Claire. I stare out the window. I want to go home. I'm in the wrong place: I am healthier than these people.

By Saturday night, I've done yoga, meditated, written my daily gratitude list. I've knit half a scarf. Six or seven times I've pressed replay on Bach's "Sleepers, Awake," the opening cut of the only CD in the common room.

And now I am out to dinner with the rest of the inmates. Every Saturday night, the director lets the "clients" test-drive their new skills in public. No wine lists, just elegant menus. It doesn't matter. We're all ordering drinks in our heads. Mine is white wine. "I want a Ketel One martini," says Jonathan.

Claire perks up. "Ketel One? That your brand? Me, too."

"Change the subject," says Jane.

Claire looks put out. "Hey, Ann, do you know why your arrival was delayed by a week?"

"Be quiet," says Jane.

"The guy in the room next to yours offed himself," says Claire. "Hanged himself from the bedpost."

"That's not possible," says Jonathan.

"Well, I don't know if that part's true. But he hanged himself all right." Claire sucks her teeth for a moment. "Said at lunch that he couldn't imagine living without drinking. I guess he was telling the truth."

"That's enough, Claire," says Jane.

"I'm just saying," says Claire, pouting at the menu.

After dinner, there are urine tests, and then two options: bed or watching *Law & Order* reruns with Charlotte. Charlotte is a self-confessed "frequent seeker of sobriety"—this is her sixth attempt at rehab—and she has established solid control of the remote. Her home group, as she's fond of repeating, is the "Church of the

Heavenly Dressed" in Manhattan. Charlotte misses her dogs. She carries photos of them in her purse: they're with her housekeeper. Charlotte loathes Claire. The feeling is mutual.

"Ann, let's play Scrabble." Claire is holding the board like a shield against her chest. I choose bed.

On Sunday we are given two choices: a trip to the mall or a visit to the crafts room. I choose the crafts room, which doubles as a gym, stacks of old magazines lining the wall beside a treadmill and yoga mats. Jane decides to join me. So does Claire, who marches ahead.

We gather construction paper, markers, blunt-nosed scissors, and a pile of magazines. My project: finishing a collage of what I value in my life for Monday morning. I have spent three decades in the magazine business, and this is what it's come to.

"I want that *Vanity Fair*," says Claire. "I saw it first."

"No, you didn't, Claire," says Jane.

"I'm getting on the treadmill," says Claire. "I'm going home fit. And you know what? You can all fuck yourselves."

I pick up a copy of *O: The Oprah Magazine* and find a picture of the Buddhist nun Pema Chödrön—"One of the wisest women living in the world!" Her face calms me. I decide to cut her out. Carefully, I snip around her youthful blond-bobbed head and choose a place for her in the left corner of my collage.

The crafts room is quiet. I leaf through the magazines until I find a double-paged spread of blue water. For just a moment, I let myself imagine it's early morning at the houseboat. The water is lapping by the window, and Jake is beside me, naked, his arms wrapped tightly around me. The otters have come back. They're raiding the minnow bucket. I can hear them splashing.

Now I hear the treadmill.

I change the tape. As I reach for the glue stick, I try to think about my son's face and his deep voice. This is what I do at the dentist's, just when the needle is going in.

Most of all, I try not to think of the last awful months. Leaving my work at McGill, leaving Montreal. Crash-landing back home in Toronto, an empty future looming before me. Getting drunk at my cousin's wake. (How did I get home?) Passing out at the Royal Ontario Museum, in my Hugo Boss dress. Nicholas confronting me about going to rehab. "You're slurring. This is obscene." Jake, awake in the night, worried. Staring at the ceiling. Gillian pouring tea for me and telling me I had to do something. Taking her bracelet off her own arm and handing it to me: my sobriety bracelet, emblazoned with "Never, never, never give up." Churchill's words, for battle.

I am doing battle. I am battling for consciousness. I want to go to bed sober. I want to remember my dreams. I want to know myself without alcohol.

Here I have a yellow room with wide pine floors, three Oriental rugs, a beautiful desk by a large window, and my own bathroom. Here I have privacy, and time to heal from my so-called progressive disease. Except it doesn't feel like a disease. I am one month sober, and I feel normal.

Everyone else here has had to detox before arriving. They read the *New York Times* in the morning, and discuss the collapse of Bear Stearns, sharing names of mutual friends working on Wall Street.

I know they are in more trouble than I am. Have I taken a sledgehammer to a flea? Perhaps. Maybe I could have just stayed home and gone to meetings? Too risky. I am determined to catch this thing before it lands me where my mother ended up. Still, I try not to think about how much the experience is costing me.

This is where my mind goes, as I cut and paste. I find a picture of bonefish in the blue Bahamian water, and I scissor around it. Kindergarten for the newly sober. It's Sunday afternoon, in rehab.

Massachusetts, Winter of 2008

I wake every morning in my yellow room, centered and happy: three weeks in and the snake of addiction is starting to retreat. I can feel it happening. I draw it constantly in our craft sessions: a snake slithering down one shoulder, a bird lighting on the other. I put this image in the middle of the collage, the one with Pema Chödrön's beaming face.

How will I keep the snake away? How will I do this on my own?

Yesterday's message from the group director: "It is inappropriate for you to have a lot of confidence when you leave here. Just decide that you will not drink. Put this on your fridge: 'Having confidence and having resolve are not the same thing.' The biggest temptation will be to check out, to numb. That is the worst place to go. Don't spend a lot of time shopping for a program: choose an approach. This is chronic. You have to talk about it."

Privately, he said to me: "I think your commitment to sobriety is pretty unshakable."

I am not so certain. I haven't run the gauntlet of summer evenings at the houseboat, of dinners out, of New Year's Eve. Of airplanes, of the Bahamas.

Our session with the good doctor from Harvard was different, as it always is. He speaks to us on a different plane. Yesterday, he quoted Henry van Dyke: *"For love is but the heart's immortal thirst / to be completely known and all forgiven."* To be completely known, and all forgiven: this is what we all wish for. The woman whose husband had her arrested in her nightie, the man who disappeared into a vodka bottle when he lost his son to suicide. We all want to start fresh. I want to start afresh, but right now I am in a holding tank: a beautiful, expensive holding tank south of the border.

Nicholas has come to visit me, and spent some time lying on

the white coverlet in my bedroom, confiding to me about his life. A breakup, a new relationship. Back to normal: we are side by side, lying in the late afternoon sun, talking about what will come next for him at the School of the Art Institute of Chicago. He is happy I am here, this boy who once turned to me for an answer to this question: "What hurts more: porcupine quills up the nose or having a baby?" And this one: "Do monkeys get periods?" This boy, who once believed he had found pirate's gold when he found spray-painted shells I had buried in the sand, believes I can stay sober. He has left, confident that I will keep to my course. I know it's a crapshoot.

I press play on Bach's "Sleepers, Awake," my daily ritual, and move to the window to check out the cars in the parking lot. Good: Terry's ancient blue BMW is there, snug close to the building. This means I will learn something today. The Buddhist assistant Terry, who says God is in the space between the in and the out breath, and the good doctor from Harvard: these are the two who teach me, here in rehab. These two speak to my hungry soul.

In the little library, I have found a copy of Stephanie Covington's book *A Woman's Way Through the Twelve Steps*. "In the process of defining our own spirituality, we may find that the spiritual language of the Steps reflects traditional Christian religious images and practices. Recovering women often struggle with the masculine language in the program and choose to substitute ideas and language that include feminine power. ... Translating the language and cultural experience of the Twelve Steps for ourselves is an important aspect of recovery." Amen, I say.

Terry, the good doctor, Covington. Beside my bed: Pema Chödrön, Bhante Gunaratana's *Mindfulness in Plain English*, Jonathan Diamond's *Narrative Means to Sober Ends*, Lisa Najavits's *Seeking Safety*, fresh diaries for more writing. I am beginning to piece together my crazy quilt of recovery, to make it my own. Bit by bit, I am emerging from the dark. In my diary, Marion Woodman's words: "Healing depends on listening with the inner ear—stopping the incessant blather, and listening. Fear keeps us chattering—fear that wells up from the past, fear of blurting out what we really fear, fear of future repercussions. It is our very fear of the future that distorts the now that could lead to a different future if we dared to be whole in the present."

The Houseboat, Summer of 2009

In my peer support group, they warn you that you'll meet people who are not the same as yourself. You never drank in the morning. You never crashed your car. You never even drove drunk. You never hid booze at the office. Endless "nevers."

You'll say these things, and you'll be right. And you'll leave.

Look for the similarities, they tell you. Don't quarrel with the details. Stay safely in the middle.

But where is the middle? In the city, an older gentleman and I share a sobriety date: November 3, 2008. That's about it. He had twenty years of sobriety when I first showed up, but he decided he could handle a drink. He lost everything, including several teeth. At first, I avoided looking at him. I found him too frightening. That changed when we began stacking chairs together. I try to keep an open mind: everyone has something to teach me.

Still, nothing has prepared me for the cellar of the Bethesda Lutheran church in downtown Kenora, Ontario. Jake drops me off, promising to return on time. I walk down the black metal steps, and find my way past the furnace room. I stir some Coffee-mate in a styrofoam cup, and try to ignore the small insects scurrying by my feet.

For more than forty-five minutes, the meeting is relentlessly grim. Finally, Ray has the floor. "It was the goldfish—the goldfish saved me," he begins. The small crowd is solemn. He continues: "I know, you've all heard about my goldfish. But that was all I had, after my wife and the kids left. And I was down on my knees, praying to be saved. Four days, I forgot to feed those goldfish. All of a sudden, I remembered them. So I got up off my knees and I fed them."

Ray looks as if he might cry. Instead, he beams. The others nod. "And of course, they all swam to the surface, because they were hungry." Everyone nods again. The whole room shares in his victory. Ray's done a good job, infusing this sad little evening with some drama. "Yep, they just did what goldfish do. They swam to the surface. And you know? I knew it was a sign."

This is what sobriety feels like. You swim to the surface, you feed yourself and your pets, and you consider it a major victory to be able to remember all this.

You brush your teeth before bed. You wash your face, without fail. You put on your pajamas, and you tuck yourself in.

There is no wine at your side. There is no wine in the fridge. There is tea, and maybe an orange, or a ginger cookie. Your dog body, the one you have dragged through misery and shame and grief, comes to expect the quiet. It looks over at you and begins to trust that you will feed it, and water it. "Let's rest," you say. You scratch its head. It walks up the stairs, climbs into bed. It wants no surprises. Welcome to the Holiday Inn of new sobriety.

"Can you move the covers, baby? I feel like Charles Atlas trying to get the truck off his chest." This is Jake, making me laugh in bed. I move the heavy Hudson Bay blankets, blow out the candle, and together, we look at the stars through the high window overhead. We sleep the sleep of the untroubled.

Mornings do not taunt me the way they used to, wagging a finger, cornering me with evidence. Mornings are sweet, benign, like dim-witted aunts. They meander without transgression, move slowly without lurching. They say grace, and smile.

Lunchtime is not much different. There are no Bloody Caesars, fringed with salt, calling my name. No jaunty little heels of Pinot Grigio nestled in the fridge by the Louisiana Hot Sauce. Just Fresca. "You know, I think I'll have a Fresca," I say, as if it were a meaningful choice. And it is. Jake says he'll have cranberry, cut with water, thanks. Another meal passes, liquor-free, as innocent as a child's birthday party, and just about as sweet.

At the cocktail hour, I lay out fine cheese and crackers, homemade guacamole and chips. Oral distractions. I read to him, a chapter each night from *Charlotte's Web*. "I can't believe no one read to you when you were kids." "There were seven of us. Why don't we read some more?" We sit beside one another on the driftwood bench, listening to E. B. White's words: "It is not often that someone comes along who is a true friend and a good writer." Jake rubs my back. Life is beyond good.

Still, there are moments that are unpredictable. I reach for some ice at the Gardners' cottage, and a bottle of Tanqueray, chilling for cocktails, tumbles out, kamikaze-like, into my hands.

The first summer, before I managed some decent sobriety, alcohol kept jumping out of cupboards and into my line of vision. Everywhere I turned, it found me. On billboards, on the pages of magazines. I felt hunted. And just when I was certain I had locked the front door, and battened down every hatch, it snuck in the back, tiptoeing in, under the guise of "grief." A sudden death, a canoeing accident, a freak drowning, and poof, a drink appeared in my hand, seemingly without warning.

It's important to know this: your alcoholism is good at disguises. It will sneak in when you're not looking, made up as the Party Girl.

You'll find a red drink ticket on a bathroom floor and actually pick it up, head to the bar, use it. Like a teenager, a thief.

That's what I've had to learn: your addiction will find a trapdoor—any damn trapdoor. That first August, it came in the guise of "saving the relationship." As in, if I just buy this little traveling six-pack of Beringer White Zinfandel, and quaff those little plastic bottles bold-facedly as the two of us cruise down the highway, we will be as happy as clams. Jokingly, I call it my Celtic Blood Disorder, but it's more serious than that. I used to say: "I was overserved." "You weren't overserved," Jake would say. And he was right. I was complicit in the whole affair: a sin of commission.

Had I known what last winter was to bring—or even the fall—I would have slipped into the water with stones in my shoes, and never dreamed of seeing summer again. That first true winter of sobriety was bracing: a rigorous marathon of determined effort. And before I made my final peace with quitting, there was a major death, a suicidal patch, and a terrific binge that reduced me to my knees. No, this is not for the faint of heart.

Toronto, Winter of 2009

Coming on Christmas and Paul Quarrington—the novelist and musician, one of Jake's best friends—is dying of lung cancer. He has just performed a wonderful solo gig, and now we're sitting at Proof the Vodka Bar on Bloor Street. All the others are drinking liquor. I just want some San Pellegrino. There is the softest look on Paul's face when he places my order. I am grateful, and totally raw: Jake is talking of moving to the West Coast. Things are heading in the wrong direction: what once felt so solid is evaporating at a reckless pace.

Toronto, March of 2010

I am having coffee with my friend Ted, who is only a few months sober. He has just broken up with his beautiful wife, and he's devastated. They have small children. "Maybe you'll get back together?" I say. "No," says Ted. "She won't forgive me for lying to her about drinking. Besides, you know what they say: You can't wake someone who's only pretending to be asleep."

I tell Ted of Jake's plans to move to Vancouver. He says: "This is not good. You know that, right?"

I am silent, but in the pit of my stomach, I know it's true. There have been some bitter quarrels on a trip to Mexico. In fact, we have had two of the worst fights of our relationship in recent months. I can feel the rumbling of an earthquake, and I don't know how to stop it. I want to live together, badly. I am tired of the long-distance arrangement.

It's an Updike sort of conundrum we're stuck in: an all-too-modern love riddle with no easy answer. Ann loves Jake; they each have a child, with ex-spouses in different cities, cities where they each have roots. Where should they settle?

Ann knows it will not be her city, Toronto—Toronto is not right for Jake. Jake proposes a commute: every two weeks, from Toronto to a third city, Vancouver. Ann is frightened: she knows this is untenable. What they do not say: Ann should move to Jake's city, Winnipeg. Ann is terrified to bring this up. She fears it is not what Jake wants to hear. She wants to be near him, always. Ann is at a total loss. Stalemate.

Toronto, April 2010

On my fiftieth birthday, Jake gave me a beautiful photo album with our history in it, with Christopher Marlowe's words inscribed in the front: "Come live with me and be my love, and we will all the pleasures prove."

We draw diagrams on napkins, trying to work out what will happen. Meanwhile, my heart is breaking. I wear our engagement ring, but I do not hear the words I want to hear: "Come live with me." I am certain I am his love.

Yesterday, I drove Jake to work and recited the one prayer I remain faithful to:

I said to my soul, be still, and wait without hope,
For hope would be hope for the wrong thing; wait without love,
For love would be love of the wrong thing. . . .

I turned my head at a stoplight, and there were tears streaming down his cheeks.

"Why are you crying?" I asked.

"I love you, baby. You have such a beautiful soul."

Toronto, Winter of 2011

No one warns you about this, but it's true: when you are stone-cold sober, the past will start sneaking up on you. All that you have drowned with alcohol will swim to the surface. Depression returns, with its old familiar face. Old anxieties appear. It's deeply dispiriting, and it confounded me daily. Jungian Marion Woodman wrote, in *Addiction to Perfection*: "One of the greatest difficulties in dealing with food addicts, as with alcoholics, is helping them to overcome their sense of despair when they lose the high associated with these addictions."

For some, new sobriety is a pink cloud. Not for me. Without a plan, cocktail hours were tough. Parties? Sometimes I had to breathe through them. It took a full two and a half years and a lot of hard work before I found my equilibrium. Once it was established, it was unshakable: I had a calm I had rarely experienced.

But by that time, Jake was gone. Eighteen months into my sober journey, he called me on a Monday morning and broke up with me on the phone. Just like that. It came out of the blue—two weeks after a particularly romantic email exchange on May 1, the day we called our anniversary: I was in New York, visiting Nicholas, and I woke to an email itemizing the "Fifteen things I love about you." It was a keeper, articulate and effusive. Among the things he loved: "The way you snuggle up against me when we sleep. Your scalpel intellect. Your inquiring, open mind. I love your courage in dealing with ideas. The smell of your hair, your neck, your sweaters. The way you hold my hand when we walk down the street." It went on in loving detail, and ended with number sixteen: "How about our long-standing date at the National Magazine Awards?" The awards were only weeks away: we already had our tickets and were discussing what dress I would wear.

Seventeen days later, the phone call came. "Surely you don't mean forever?" I said. "Maybe we could try again after Labor Day," he said. He paused. "No, I take that back." I couldn't make the news go into my head. "Let's talk midweek," he said. I called. He would not pick up the phone. I did not call again.

Perhaps there's no good way to break up with someone—but this was a parting that made no sense. Not to me. Not to many around us.

Maybe this is the way it always transpires: one person leaves, and the other is shattered. Decades ago, I broke up my marriage to Will, and he spent weeks trying to talk me out of it. I would not budge.

All I know is this: my breakup with Jake still makes no sense to me. His absence is a presence for me, one that informs my writing, my thinking, my way of looking at life. I've grown used to it, as you do a much beloved dog: it accompanies me wherever I go. The writer

Edward Hoagland once called his marriage breakup "the rip in his life." This was—and remains—the major rip in mine. I miss Jake each and every day. And yes, I still love him with my whole heart.

Twenty-four hours after Jake's call, numb with shock, I learned that I had won the prestigious Atkinson Fellowship in Public Policy, charged with delving into the subject of women and alcohol: a yearlong project that came with a generous travel budget. After three years of staying at home to heal, I was back in business.

Still, I was bereft: that summer was the toughest of my life. Eight weeks into the breakup, Canada's national newspaper ran a piece exposing the details of our relationship and what had happened. Written by a mutual friend, it posed this question: why had we spent fourteen years in two different cities, never resolving our living situation? Pseudonyms were used, but it was no mystery as to who the couple was. It called me smart and sexy, but this was cold comfort. As the husband of a friend observed, "Names were changed to protect the guilty." My titanic heartbreak had a very public airing.

Less than six months after Jake left, my beloved father was dead. When the doctor called to explain, he asked me if I was aware of Korsakoff's syndrome. With a quiet voice, I said yes. When he asked me if I knew what a twelve-step group was, I said yes again. He suggested Al-Anon. I smiled. It was all too Gothic. My beautiful father—stoic, singular, brilliant—had fallen into the trap. I thanked the doctor for his kindness and hung up the phone. Later, my sister would find an Al-Anon card tucked away in my father's filing cabinet. I found this unspeakably sad. It all is.

Breaking the Trauma Cycle

THE MOTHER-AND-CHILD REUNION

The past is never dead. It's not even past.
—WILLIAM FAULKNER

At two, she was taken from her mother by her father. She lived in a truck for a year, with his mother and his sister. At seven, she started sipping beer and was taken from her father—"just say it was sexual abuse." She moved in with her aunt and uncle. At eight, she went into the first of many foster homes. At eleven, she moved to a group home. At twelve, she quit school and began "living on the street." This, I learn, is a euphemism for prostitution. At thirteen, she was pregnant. At fourteen, she gave birth to the first of her four children.

Now thirty-two, Annie Akavak can tell her story without reaching for Kleenex—a change from when I met her at a treatment center two years ago. "Alcohol was my first experience of getting high," she says. "I remember wondering, 'What can I forget? What else will numb the pain?'"

From alcohol, Akavak moved on to marijuana, then ecstasy. Eventually she used crack. "Crack takes all the pain away. It numbs everything. The only thing I never did was shoot up—I'm afraid of needles."

Today Akavak is in the process of putting her life back together again. It has been a very long road. Four years ago, her newborn daughter was taken by Native Child and Family Services of Toronto. "They don't mess around," says Akavak. "I swore that I would fight to get her back."

She did, with the help of the Jean Tweed Centre in the west end of Toronto. Enter the doorway off Evans Avenue, under the humble striped awning, and you confront a world shaped by Nancy Bradley and her team. Treating substance use is the focus of the center, whether inpatient or outpatient. Clients include sex-trade workers and university students, women from the corporate world and those with a criminal past. Bradley, who has been at the helm of Jean Tweed for twenty-five years, has seen them all: "a microcosm of society." "Twenty years ago, we made some mistakes," she says. "We used to believe that you dealt with the addiction and told the woman to wait two years to deal with her other problems. Now trauma pervades everything we do. You can't separate recovery and trauma. Women may not be drinking, but they will still be very, very troubled if you don't address the underlying issues."

"Braiding" is the approach used at Jean Tweed: the ability to move back and forth between a woman's addiction and her other challenges. It might be violence, it might be sexual abuse, extreme poverty, or a traumatic childhood. Rare is the woman without a troubling past or present. "Someone can't go through substance abuse who isn't anxious, depressed, having issues related to self-harm or perhaps food," says Bradley. "We meet women where they are. We know that some behaviors may manifest themselves. Some may begin to have flashbacks—which can be extremely painful for both the clients and the staff. We try to help them see the links. We hear some of the most unbelievable stories—there is immense courage here."

How one deals with trauma is key. "Trauma shuts down the frontal lobe," says Susan Raphael, a Toronto counselor who specializes in dealing with young clients. "If you don't deal with the trauma, it recycles. But you have to move away from the narrative: make the connections to the past, but not have the client relive it." Often, being addicted to substances means that the woman will be retraumatized by experiences that happen when they're under the influence. "Young women still really objectify themselves," says Raphael. "I would have thought things would have changed by now. But I see clients flirting with being escorts, having a sugar daddy—what I would call a watered-down sex trade."

"We miss the biggest part of the story if we don't link the addiction to the rest of the woman's life." This is the voice of Nancy Poole, director of research and knowledge translation at the British Columbia Centre of Excellence for Women's Health. With more than thirty years' experience in the field of addictions, Poole is a dynamo, with her finger in dozens of projects. She is talking about the importance of what is known as trauma-informed care, which gives credence to the woman's past or present, and the role it plays in her addiction. Co-editor of a new book called *Becoming Trauma-Informed*, Poole says: "We are neophytes when it comes to dealing with trauma. The term 'trauma-informed care' has some cachet, but people have very little experience in translating it. The way we have been delivering care has been retraumatizing—crippling instead of empowering. It's really about our collective denial about child abuse and violence against women. Trauma is quite common for people, and it can interfere deeply with a person's ability to cope. Trauma-informed care is supportive, creating a safe place—the Jean Tweed Centre, for instance, is *years* ahead of others in Canada."

Head to one of the most renowned treatment facilities in the world—the Betty Ford Center in Rancho Mirage, California—and you will hear a similar story. Johanna O'Flaherty, vice president of treatment services, stresses the correlation between trauma and addiction. "A traumatic event is a dramatic event that is so extraordinary that it's outside the individual's coping abilities," says O'Flaherty. "It does not toughen the child or the individual, but it toughens their defenses. Defenses serve you very well. Drugs and alcohol may actually be a wonderful anesthesia to keep the pain numbed. But once one stops the drugs and alcohol, the individual will reexperience the painful feelings. And there is now research that validates that if trauma is not addressed, there is a high propensity for relapse in the first six months. It's all about moving from 'victim' to 'survivor.'"

O'Flaherty stresses that not everyone who ends up being an alcoholic has early childhood trauma, and not everyone who has trauma ends up being an addict. But she does underscore the importance of this work, which she has brought to the Betty Ford Center in the past six years. Beyond trauma, what is the largest issue she is confronting at Betty Ford? She doesn't blink. Mixing other drugs with alcohol is the first thing she mentions—"benzodiazepines and liberal prescriptions of Ambien." Says O'Flaherty, "There are enormous barriers for women accessing treatment. What's really lacking in the field? Treatment centers that will accommodate babies."

At the Jean Tweed Centre, helping clients change their lives often includes assisting mothers with their parenting skills or custody problems. Twenty-four years ago, a young woman barged into Bradley's office with her seven-year-old, desperate for help. She said, "I have to give my kid up to Children's Aid—he is out of control." The young woman had been through Jean Tweed and felt a part of the place. "I thought to myself: 'We have an opportunity here,'" says Bradley.

With that, Jean Tweed began offering child care so that parents could attend evening sessions. Twelve years ago, they started Pathways to Healthy Families, an outreach program aimed at helping women who are pregnant or parenting children up to the age of six. They placed staff outreach workers in shelters, in a maternity hospital, in the aboriginal community: looking for women who needed help finding housing, prenatal care, midwifery, counseling for substance abuse, and more. They built bridges with agencies that were historically adversaries: Children's Aid Society of Toronto, the court system.

As part of the Pathways program, they developed the Mom & Kids Too program, tailoring the essence of the classic twenty-one-day treatment program to a young mother's schedule. Treatment is spread over seven weeks, three days a week, with child care in their licensed facility. Workshops on parenting, play routines, and attachment are incorporated, along with sessions on substance abuse. In the morning, mothers and their children can arrive and have breakfast together. With this program, retention of young mothers went from 26 percent to higher than 80 percent. Says Bradley, "Often, women haven't been parented well themselves, so we're role-modeling."

Akavak used both programs to help her get her life back in order. When her daughter was taken from her, she found Pathways and Jean Tweed, enrolling in Mom & Kids Too as well. "That's when everything changed," says Akavak. "I stopped relapsing, and started learning some coping skills." Gradually, Native and Family Services allowed her to have her daughter with her for the day sessions. Her daughter was in foster care. "They were seeking adoption for her at the time," says Akavak, "with no opportunities for visitation. I wanted her back—I didn't want her to feel that she was being pushed away as I was as a child. I wanted to be there."

Over time, with Jean Tweed's help, Akavak won her fight for custody. "They saved my family and they saved me," she says. "To have someone acknowledge that you are willing to make a change is everything. Jean Tweed taught me how to be a better mother, and a healthy mother. I know I am going to be okay because I'm going to make it different for my kids. That's what makes me strive to be strong."

Today, Akavak is back at school. Her two youngest children are in day care, she is doing two kinds of therapy, and she can see a path forward. I ask her if she is in a relationship, and she shakes her head. "I have to love myself before I can love anyone else," she says. "It feels like it's going slow, but there's a lot of self-healing. Self-healing, self-love, self-forgiveness. That's the biggest thing for me."

Jerry Moe, the founder and head of the Children's Program at the Betty Ford Center, is a clone of William Macy: an open-faced, open-hearted guy who gets right to the point about "his best teachers"—namely, kids. And since his clients are pint-sized, he spends a lot of his day on his knees, speaking at their level, helping children separate the person they love from the disease that's consuming them.

"The Children's Program is about breaking the cycle," says Moe. "Roughly seventy-five percent of the people in treatment come from families with addiction. My role is to teach kids self-care in a family that doesn't practice self-care—this is a strength-based program. And sometimes the first person in the family to get help might be the seven-year-old.

"So many of the kids are 'looking good' kids—most are suffering in silence. These kids are guarded: they don't want anyone to know the family secrets. Remember, this is a disease of silence and secrecy. They may not know the family problem is drinking, but they know something is wrong." How long does it take for a child to open up to him, to build trust? "Typically, about an hour," says Moe. "No, actually, thirty minutes."

Moe works with children ages seven to twelve, doing "all sorts of experiential activities that help kids learn to thaw all those emotions that are kept in check. We're with them for four days," he says. "Kids can't look good all day. When you're with them all day, you really get to see how they deal with frustration, what they do when they're sad. It's a great opportunity to build relationship, to get a window into the way kids operate. At some point I will say, 'All of us have someone we love who got trapped by addiction. Tell on the disease, on addiction. It loves when we are silent.'"

One of the activities Moe relies on is drawing: "Kids can often draw what they can't say." Another centers around a backpack full of rocks: about forty-one pounds in all. "Each of the kids gets to carry it briefly," he says. "We talk about what our lives would be like if we had to carry it all the time. 'I'd be miserable,' kids will say, or 'I couldn't ride my skateboard.' They'd do anything not to carry it. We say: 'Many of your moms and dads are carrying that backpack. Where do you think they carry it?' To which one seven-year-old girl replied: 'It's somewhere in that space between your heart and spirit.'

"How long do you suppose they've been carrying it?" asks Moe. "Since they were little kids," he says. "So how can anything in that backpack relate to anything you've done?"

"Whatever's in that bag, why don't they let it out?" asked one young girl. "They don't know how," Moe tells them. "If they do drugs or drink, it goes away. But when they pick it up again, it weighs more. It gets to the point when the bag hurts all the time. That's when they go to treatment."

All the rocks are painted. One reads "secrets," one says "fighting," three are named "problems"; one says "addiction," another says "abuse." "There are four kinds of child abuse," says Moe. "They typically talk about physical abuse, maybe verbal abuse, occasionally someone will talk about sexual abuse. Then there's neglect.

"While we're doing this exercise, we're still developing intimacy, informing them and helping them understand that people can get it out or keep it in. How do people get better? They talk about a secret. And by the third day of the four-day program, we teach the children that they have their own bag of rocks."

"Part of this is breaking the belief that it is scary to talk about," says Peggy McGillicuddy, a counselor who has worked for years with Moe. "There's not much denial with little kids. Once one kid talks, it's amazing what the other ones will share. Parents are always so scared—and it's always the opposite. They just love their parents.

"On the second day, the kids write their own true story. 'Who in your family got trapped by addiction?' On the third day, the child reads the story. 'Tell your mom what you were worried about,' I say. A lot of them will say: 'I was worried you were going to die.' Kids are allowed to be angry at the disease and still love their parent. Kids get to talk about how awesome their parents are—even if they aren't getting sober."

McGillicuddy herself is the child of a female alcoholic—Mary Gordon, who just happens to be head of the Family Program at Betty Ford. On the week that I visited the Betty Ford Center, there was a full complement of families attending the program, including one from the United Kingdom and one from Saudi Arabia. "This is the week when we disturb the comfortable and comfort the disturbed," says Gordon. "These families may have been functioning, but they have been functioning in shame. We have a resiliency model—we try to avoid the word *dysfunctional*. Of course, we use the three C's—you didn't *cause* it, you can't *cure* it, you can't *control* it. But you can cope in new ways. Ideally, the women learn about the self-care piece, and the men learn how to let go of the fixer or problem-solver roles. They will learn not to react, rescue, caretake."

Betty Ford has every right to be proud of these programs. It's worth noting that in the past decade, many children's programs have disappeared. Says Moe: "So often in our field, these programs are the first to be cut. If I had a dream? All treatment centers would have to have a children's program to become accredited."

What does it feel like to be a mother in recovery, helping young children make sense of what they remember of your drinking past, and to adjust to your sober present? A woman I will call Lesley agrees to meet me for coffee, and to discuss why she is determined that her children will be exposed to "others in similar circumstances, to help them to talk about alcohol openly and realize it isn't a big secret. To draw it out is not an easy process—the feelings and emotions."

Poised and beautiful, with a mellifluous voice, Lesley is a picture of self-control. Her story is otherwise. The forty-two-year-old mother of two says: "I loved drinking from the very beginning: the sense of ease and comfort it brought. It made me feel confident. It was easy to socialize. But at sixteen, I blacked out with my first drink. My family drank heavily, and no one criticized me when I would come home from a date and pass out on the kitchen floor. And by the time I was eighteen, I was starting to feel the shame. I knew it was causing me heartache: all my relationships were falling apart. And for that reason, I could not wait to leave home."

At twenty, Lesley moved to Toronto—but the geographic cure did not work. On her first night in the city, she went to a downtown pub, and ended up falling in a ditch. "That was the story of my life," she says. "People having to get me home."

Eventually, Lesley fell in love, married, and had two children, now twelve and eight. Married life came with a family cottage—a lifestyle that was suited, as she says, "to an alcoholic. Caesars at breakfast, beer in the afternoon, wine at dinner, then shooters of tequila. I loved it—and I hated it." Eventually she decided she should cut back her drinking—and she spent several years trying to find a solution: "I did a program of controlled drinking. That didn't work. I couldn't stop. I did hypnosis with a doctor. That didn't work. I tried drinking something I didn't like—beer. That didn't work. I asked my husband to tell me when I had had enough. That didn't work because I would say, 'Don't tell me what to do.'"

Her drinking caught up with her. One afternoon, she left her two young children and three others she was babysitting to buy vodka. That night at dinner, her young daughter described what had happened. "No, I was just moving the car," said Lesley. Her daughter challenged her story. Lesley had to tell the truth.

Within weeks, her siblings intervened on her at a family reunion. By that point, says Lesley, "I was destroyed, emotionally, spiritually, physically destroyed." She joined Alcoholics Anonymous, and she stayed. She found a new job, working with people she likes. However, her marriage ended and there has been a lot of "emotional turmoil" for her children.

Most recently, Lesley has had her daughter and son take part in the Children's Program at Renascent treatment center in Toronto. "There was a lot of turmoil," she says. "I was very worried about my children's emotional well-being."

Heather Amisson, a family counselor at Renascent who leads the adult portion of the Children's Program, is herself a mother in recovery—a wife of an alcoholic, a daughter of two alcoholics, a mother of two, a niece, and an aunt. "It's really difficult for us to forgive ourselves," says Amisson, who has been sober for many years. "Even now, it's a struggle for me." In the Children's Program, each child writes a letter to the addiction of their parent. They also make a family shield, one that articulates the family traditions that would make a difference in their life: meals together, shared outings. The shield is then framed and presented to the parent. "It's very power-ful," says Amisson. "We really believe addiction is a family disease, and often the addicted person didn't do the simple care things. This weekend brings a recognition that traditions matter. Meanwhile, it's really important that we present things in a way that doesn't cause the mother to shut down. The shame can cause a woman to do this—feelings of remorse, guilt, living in the past. We're trying to keep the shame and blame off, and help them take care of their family."

For Lesley, this has meant establishing new traditions with her children. Friday night is pizza and movie night; Sunday means going to church and then to a local bookstore. They play charades on Sunday night. Says Lesley, "It's not an easy process. But I want them to feel loved." Today she is not ashamed of her drinking. "It's a victory," says Lesley. "I fought a battle that not many win. There is hope."

15.

Something in the Water

SHAPING A STRONG PUBLIC HEALTH STRATEGY

Here's a question. Let's say there's a frog pond where a growing
number of frogs are developing odd-looking growths, and
others are becoming sterile. Do you send in surgeons to remove
the growths, and fertility experts to deal with the sterility? Or
do you say to yourself: maybe there's something in the water?

—DAN REIST, ASSISTANT DIRECTOR, KNOWLEDGE EXCHANGE,
CENTRE FOR ADDICTIONS RESEARCH OF BRITISH COLUMBIA

Is there something in the water? Of course there is. We live in an alcogenic culture, one where risky drinking has been normalized. We
swim in an ocean of advertising, and that advertising says one thing:
drink, and great things will happen. We absorb this in our pores. In
fact, it's so prevalent, we barely notice it.

Robert Brewer knows there's something in the water. "This is a
huge public health problem," says the leader of the alcohol program
at the National Center for Chronic Disease Prevention in Atlanta,
a division of the Centers for Disease Control. "We have to broaden

our understanding of what we consider an alcohol problem to be—well over eighty percent of frequent binge drinkers are not alcohol dependent."

Robert Strang agrees. The chief public health officer of the province of Nova Scotia, Strang knows that the culture of normalized heavy drinking is growing. "This is not an addiction issue," says Strang. "Addiction is the far end of the spectrum. This is about the impact of alcohol right across society. Lots of harms are coming from those who are not addicted. Periodic, episodic binge drinking leads to acute and chronic problems in society. The problem with alcohol? We don't acknowledge it as a drug—and as such, we haven't paid enough attention to it.

"It's about changing social norms," says Strang, "getting those communities already aware of the damage to work together—the medical community, the FASD community, the violence against women community, the road safety community, the breast cancer people. We need to have a robust discussion about this issue: How does alcohol play out in your community? In terms of suicides? Kids being abused? Violence? Teens in emergency rooms? Are we having an adult discussion? I don't think so."

Is there something in the water? Mike Daube knows there is. Well seasoned from the tobacco wars—he was the first full-time director of Action on Smoking and Health in the U.K.—Daube is cochair of Australia's National Alliance on Alcohol. Says Daube: "Alcohol is where tobacco was forty years ago. And it's the same: a massive and powerful industry, cynically promoting its product. There is a ruthless recognition by industry that young people have more access, freedom, and money than ever before—products are designed and marketed, targeting young women to get drunk as quickly as possible."

There is no doubt: the U.S. Congress, state legislatures, local city councils, and provincial governments around the world are in the pocket of big alcohol. The alcohol industry employs one lobbyist for every two members of Congress. It gives generously and across political party lines. And it succeeds, over and over again, at blocking evidence-based public health steps to control alcohol problems.

Is alcohol the new tobacco? Strang believes the answer is yes. A veteran of the tobacco fight, he is determined that the harm from drinking be recognized faster than it was with smoking. "We had to work for thirty or forty years on tobacco," he says. "If we apply what we learned on tobacco control concerning price, advertising, and access, we could make significant progress on alcohol in a much shorter period of time."

What Brewer and Strang and Daube are envisioning is a comprehensive public health response to the harm caused by alcohol. They are far from alone in their fight to move alcohol to the top of the public agenda. But what government wants to tamper with our favorite drug? Says alcohol policy guru Robin Room, who has experience in the United States, Canada, Sweden, and Australia: "As market-friendly governments get more desperate as to what they're going to do about alcohol, you see a move back into a more individualized control system: deal with the bad-apple killer drunk and leave the market alone."

Market-friendly governments may want to ignore the broader picture, but the evidence is building: alcohol-related harm is widespread. It's costly. It's disturbing. This is a public health issue, and it's begging for leadership. Says Jürgen Rehm, director of social and epidemiological research at the Centre for Addiction and Mental Health, author of more than five hundred journal articles and ten books: "When you consider the science, alcohol is doing the most harm in our society. Unless we start seeing leadership on alcohol policy, our life expectancy will decrease. We should move on taxes, on pricing, on advertising, on the general availability of alcohol."

What is public policy? At its very essence, it's a simple equation: evidence plus values plus politics equals policy. We have solid evidence that widespread risky drinking is costly. We also have solid evidence that key policy levers will influence this picture: upping price, restricting the accessibility of alcohol, and limiting marketing are the three strongest. Press on those levers and you can shift consumption. You can also tackle some other big problems along the way: alcohol taxes are easy to administer, and would cover many shortfalls in cash-strapped times.

But society has to want to make those shifts—and to do so, we have to be clear about our values. Let's face it: when it comes to alcohol, our values are a little fuzzy. We tend to *other* the problem: it's the rare alcoholic, the drunk driver, the guy on the street corner swigging from the brown paper bag. And if it's our own problem? Well, we're just trying to drink like the French or the Italians.

Actually, only a small proportion of the population are alcoholic. If you eliminated all risky drinking, you would decimate alcohol sales. Says Rehm: "Our drinking patterns are not benign. Alcohol consumption creates more harm to others than secondhand smoke. It's about time we took a hard look at the problems that drinkers cause in their immediate environment and in society at large. This starts with family problems and ends with drunk drivers." In 2010, in the first major study of its kind, Australian researchers estimated that the costs of harm to others matched the traditional costs of the drinker to society.

Solid evidence plus fuzzy values adds up to political inaction—inaction for which women may pay a bigger price than men. Policy isn't gendered. No, says Sally Casswell, director of the Centre for Social and Health Outcomes Research and Evaluation at Massey University in New Zealand. "But what we're looking to do is to prevent the early uptake and heavy drinking of young women—and we can only do this through state policy versus educating them."

Tim Naimi, a physician and alcohol research scientist at Boston Medical Center, takes it one step further. "Parity in alcohol consumption is one area in which the otherwise wonderful overall trend towards greater gender 'equality' is shaping up to be an unmitigated disaster for women. Women arguably have the most to gain from a strong policy environment: they are disproportionately likely to suffer the health and social effects of alcohol abuse, to suffer from interpersonal violence including sexual assaults, and sexually transmitted infections. They are definitely prone to the secondhand effects of excessive consumption."

Policy shifts can cause seismic cultural and environmental shifts. This is well documented. Look at drunk driving. Look at tobacco control. But as the feisty, seasoned California activist James Mosher says: "With tobacco, it was easier. We eradicated the product. With alcohol, you want to manage it. But as you know, politics are so driven by special interests—and money. We don't have the power base to counter the strategic lobbying of the alcohol industry."

Which raises the question: what would a comprehensive public health response look like? Since alcohol-related harm is an under-the-radar issue, each country should begin by creating a comprehensive national alcohol strategy. Strategies need teeth: strong federal endorsement, with the understanding that all levels of government will engage in the solutions, including regional and municipal, state and local. Each has a role to play.

Strong strategy should make a priority of the three most effective policy tools: taxing and pricing, restrictions on accessibility, and limitations on marketing.

When it comes to taxing and pricing, the implementation of minimum pricing and the indexing of price to inflation are key. The following formula is time-tested and true: price goes up, consumption goes down. Minimum pricing is the most targeted approach because it has the strongest effect. It targets the youngest drinkers, who prefer cheap alcohol, and the heaviest drinkers. It's the most feasible and publicly acceptable because it's not hitting everyone's hip pocket. And the health benefits are significant: many fewer people die.

On pricing, British-born Tim Stockwell is the international go-to guy: this is the man the Scottish parliament turned to when they wanted advice in 2011. Currently head of the Centre for Addictions Research of B.C., Stockwell gives Canada full marks for its minimum pricing: "This country is almost alone in setting the floor, or minimum prices for alcohol." In a recent paper in the *American Journal of Public Health*, he reports that a 10 percent rise in the price of alcohol is associated with a 9 percent drop in hospital admissions for acute alcohol-related issues—"those people getting drunk and injuring themselves"—and a similar drop over two to three years in admissions for chronic alcohol-related problems.

Scotland has indeed declared its intention to set a minimum price for a standard unit of alcohol at fifty pence—a decision backed by the medical profession. There was an immediate backlash. The policy was challenged by the European Union: backed by such wine-producing countries as France, Spain, and Italy, the EU said the minimum pricing breaches free trade. Moreover, the Scotch Whisky Association and Spirits Europe were granted judicial review of the legislation: they have many arguments, including one saying that minimum pricing would damage a valuable export industry. Whiskey is Scotland's number one export after oil and gas. The Scottish government won the first round in what looks to be a lengthy court battle when a judge ruled that its decision to set a minimum price for alcohol was legal and justified. The brouhaha pleases Stockwell, whose research has been at the center of much of the press: "There hasn't been such a sustained debate about alcohol policy in many, many years—something good will come of this."

Meanwhile, Prime Minister David Cameron seems to have reneged on his promise to stop the flow of cheap alcohol being sold in British supermarkets. He has been quoted as saying that he wished to eradicate the reality of twenty-pence cans of lager. A base price of forty-five pence per unit had been proposed for England and Wales, but Cameron seems to have abandoned his commitment to minimum pricing. The medical community and others are pushing him to be courageous: "don't wimp out" is the general message. In the U.K., there are more than a million alcohol-related hospital admissions each year.

Very few expect the United States to tackle minimum pricing. In most states, alcohol is just plain cheap. You can visit a corner store in many American cities and find that beer is the cheapest liquid available—cheaper than water, orange juice, or milk. Federal alcohol taxes have not been raised since 1991. Historically, there was a time when alcohol taxes accounted for 30 percent or more of federal government revenues; now they account for less than half a percent.

Few states have addressed the tax issue. The most recent hike was in Maryland in 2011: significant politically, but minimal in price—three cents on the dollar. Bear in mind that in the United States, alcohol-related costs amount to an estimated $220 billion annually. Ideally, Congress would maximize excise taxes, and link them to inflation. This could have a major impact and raise revenue—but there is no sign of anything happening in the foreseeable future. "The landscape is pretty bleak at the moment," says Naimi. "Ultimately, cheap booze is not healthy for society. And industry is framing the issue—not public health."

Says the influential Thomas Greenfield, director of the California-based Alcohol Research Group: "Public health won the battle against tobacco. But in the alcohol area, the producers are shrewd. And gender convergence is propping up sales. The cocktail culture is alive and well."

Which brings the discussion to accessibility. This one's simple. "The fewer miles between you and the nearest alcohol outlet, the more likely you are to drink," says Stockwell. "If you let the market control alcohol access, public health and safety are out the window." In much of the United States, it's possible to buy beer and wine 24/7 in grocery and drugstores. In Britain, alcohol is often used as a loss leader in its four major supermarket chains: Tesco, Sainsbury's, Asda, and Morrisons. There alcohol is 44 percent more affordable, in real terms, than it was in 1980. Accessible, and cheap.

When it comes to marketing, I turn to David Jernigan, director of the Center on Alcohol Marketing and Youth at Johns Hopkins University. He points out that after decades of using women to sell alcohol to men, alcohol companies have discovered they need women to drink—and they are marketing accordingly. And countries like the United States, with its protections for "commercial speech," offer a clear runway for that marketing to take off. "The solution is going to be different in each setting," he says. He cites South Africa, where there is a proposal to ban all alcohol advertising. "In South Africa, drinkers tend to drink heavily, but most of the population abstains. Advertising will recruit new drinkers into that heavy-drinking pattern. With the heavy problematic drinking that exists, allowing advertising a free hand is a mistake."

One example Jernigan supports: "France's Loi Évin, which is a just-the-facts approach to advertising, without lifestyle features." France passed this law, named for its then minister of health, Claude Évin, in 1991: advertisers may not use images of drinkers or depict a drinking atmosphere. Only the product is advertised—a refreshing approach.

In Canada, the national broadcast regulator used to review all alcohol ads. Since 1997, guidelines have been essentially voluntary. However, much of what is seen by young people does not qualify as traditional advertising. And there is no policing of YouTube or Facebook. Says Jernigan: "Social media is all about engagement, and it's all unregulated. Marketing through social media is so cheap, and it's so targeted. Tobacco companies have been forced to restrict their activities there—the alcohol industry should take a page out of the tobacco books before someone else takes the page out for them. If it's going to police itself, then police itself. Right now, the industry has set the agenda, and it's a race to the bottom. It's public health's job to set the floor. At a minimum, I would say to the industry: do no harm. I would ask them to refrain from digital marketing, and I would not let it be voluntary. We have a natural experiment happening—and it's gendered."

Beyond the three key levers of change, there are many other endeavors that could have an impact on how we view and consume alcohol. One excellent idea: annual report cards at the national, state, and local levels, using common indicators to measure acute and chronic alcohol-related harm: alcohol-related assaults, emergency room admissions, morbidity, family violence rates, impaired driving charges, outlet density per capita, and more. Jurisdictions need to monitor the impact of harm to others. Taking this tally is key. "FAS or FASD is just one example of harm to others," says Rehm. "Drunk driving—which is typically a male problem—is another. Ethically, morally, the point of both is that no matter your politics, there is a certain responsibility concerning your impact on others. There is no freedom or right that an individual has that will excuse harm to others. If you agree with that, it follows that any kind of measurement is a top policy item."

With respect to treatment, countries need to fully endorse and invest in an integrated strategy. In North America, only 10 percent of those with alcohol dependence issues end up getting treatment help. In Europe, the figure is closer to 8 percent. Of course, not everyone with an alcohol problem needs treatment: Rehm estimates that 40 percent would be ideal. We need significant, targeted reinvestment in addiction services, addressing the vast gaps in availability and accessibility, with special attention to isolated, rural, and remote regions and vulnerable populations. Primary care needs to do its part, but that's just part of the puzzle. A broad spectrum of tiered and networked services is imperative—and an excellent investment. "The potential is there now, with the Affordable Care Act," says Susan E. Foster, vice president and director of policy research at the National Center on Addiction and Substance Abuse at Columbia University in New York. "We know that two-thirds of people in prison meet the criteria for addiction, and only eleven percent in the criminal justice system get any sort of care. There is a giant disconnect here. Addiction is not a moral failure—in most cases, it's a developmental disorder. It's a complex brain disease, and it's treatable."

At the same time, we need to expand each nation's capacity to do screening and brief intervention. In the past two decades, studies have shown that early intervention yields extremely beneficial results. Systematic access to screening and brief motivational interviews should extend to all emergency departments, university health clinics, public health sites, and other points of contact. Short in duration, these can be handled by a primary care practitioner, a nurse, a social worker, or any allied health professional. The challenge? Pressed for time, many frontline doctors feel too rushed or ill-equipped to screen for alcohol.

On the drunk-driving front, random breath testing is the future. Says Andrew Murie, CEO of Mothers Against Drunk Driving Canada: "It's largely a male crime—females make the mistake of agreeing to be passengers. They could be a major influencer on male behavior." In 2009, 38 percent of all Canadian road deaths were alcohol-related. In Sweden—which has random breath testing, and where the legal blood alcohol limit is .02 percent—the percentage of road deaths due to impairment sits in the low 20s. Says Murie, "Canada and the United States need to catch up with the world." Practiced in Australia, New Zealand, and many European countries, random breath testing—or compulsory breath testing—means that drivers are required to take a preliminary breath test, even if there is no suspicion of an offense. "It's now been more than five years since Ireland adopted random breath testing," says Murie. "And the amazing thing is: deaths due to impaired driving are down fifty percent since it came into law in 2006.

People have got the message, and changed behavior. That's an incredible story: it saved lives—as well as valuable judicial resources and money in the health-care system. Australia has had random breath testing for thirty-plus years, and they have maintained those savings. They just shake their heads at us."

Finally, we need to invest in sex-specific, gendered research. We're just beginning to understand how differently women respond to alcohol. Historically, women have consumed less alcohol than men, and for that reason, we have considered them less at risk. Alcohol affects women's bodies differently. We need to use sex-specific criteria for calculating its effect. We need to reframe the research agenda on this subject. Says Rehm, "We do seem to have losers and winners in the trends, and women seem to be the losers in alcohol-related harms."

We know that alcohol affects women's brains differently than men's. We must continue to pursue sex-specific biological research. Says Nancy Poole, director of research and knowledge translation at the British Columbia Centre of Excellence for Women's Health, "One of the things we need to understand is the coexistence of depression and drinking in women: how drinking and depression fit together—we know they are linked. How do tobacco and benzodiazepines fit with drinking? What is the integrated conversation we should be having with health providers? My key concern is: how do we understand what is facing women at higher risk—aboriginal women with high rates of poverty, violence, many of them young mothers? We need more community-based research."

At the same time, we need research to focus on the high rates of risky drinking for underage girls. "We need to understand what supports are needed to delay their uptake of alcohol use," says Poole. "Some international research shows interventions that may be helpful: ones that address such protective factors as computer-based interventions, all-girl groups, and programming that helps girls critically analyze media messages." Needless to say, there is a role for advertising and social media sending a different message. Take, for instance, a memorable Australian ad you can find on YouTube, called "Becky's Not Drinking Tonight": a young girl texts "Naah, not drinking tonight"—and it goes viral around the world—to the United Nations, to space stations, and more. Brilliant.

One of the major questions emerging globally is this: with alcohol companies now targeting female drinkers in the developing world, will better-educated women in those countries follow the path of those in the developed world? Women in developing countries are undergoing rapid social and gender-role change. Sharon Wilsnack, who oversees the forty-one-country GENACIS project, has strong views on this subject: "When I get up on my soapbox, I would like to do an experiment in those countries on the cusp. I would like to find ways of using social marketing, to help women understand that if they choose to use alcohol, they should use it in a much less risky way—and demonstrate gender superiority. Of course, excessive drinking has come to represent gender equality, and we need to turn it around. The highest risk is for the higher-educated women in lower-resourced countries. We need to design targeted intervention and tie it into empowerment."

This is a huge challenge, one the developed world shows no signs of embracing. But if women in developing countries end up enjoying alcohol in a healthier, low-risk manner, they will have pulled off something quite remarkable: a standard to which many women in many countries could aspire.

Let's dream for a minute, and suppose that both men and women adopted a less risky manner of drinking. If this were to happen, the alcohol industry would stand to lose close to half its market. Which is exactly why the alcohol industry lobbies for the status quo. Deep pockets can purchase plenty of political silence—silence that is proving all too costly on the public health front. It's time for change.

16.

Wrestling with the God Thing

SPIRITUALITY AND SOBRIETY

The spiritual life is not a theory. *We have to live it....*

If we are painstaking about this phase of our development, we will be amazed before we are half way through. We are going to know a new freedom and a new happiness. We will not regret the past nor wish to shut the door on it. We will comprehend the word *serenity* and we will know peace. No matter how far down the scale we have gone, we will see how our experience can benefit others. That feeling of uselessness and self-pity will disappear. We will lose interest in selfish things and gain interest in our fellows. Self-seeking will slip away. Our whole attitude and outlook upon life will change. Fear of people and of economic insecurity will leave us. We will intuitively know how to handle situations which used to baffle us. We will suddenly realize that God is doing for us what we could not do for ourselves.

Are these extravagant promises? We think not. They are being fulfilled among us—sometimes quickly, sometimes slowly.

They will always materialize if we work for them.

—"THE PROMISES" OF ALCOHOLICS ANONYMOUS

Shelburne Falls, Spring of 1994

A ten-day silent Vipassana retreat in rural Massachusetts—my "hair-shirt holiday." Rising at four thirty every morning, meditating for more than nine hours each day. Not only is talking forbidden; we are encouraged not to meet the eyes of others. Gillian has come with me: she is a veteran, understands what we are taking on. I do not. On the first day, I pray to God: I want a sign that I will find the courage to go forward. I need to deal with the deep depression that dogs me, and the loneliness, too.

Here, it is unspeakably difficult, and deeply rewarding. Day Four: a breakthrough. Day Six: another.

Day Ten: my last day, the day we break the silence. I take a walk at dawn to say good-bye to the property, wandering into the small field. I look up in the sky: two jet trails have crossed, making the wide horizontal X of St. George. At home, I have been reading the story of St. George and the dragon to Nicholas. On his flag, the wide horizontal X cross. As I walk into breakfast, I notice that my steps across the field—marked in the dew—have created the same image: the wide X of courage.

I take it as a sign: what is in the sky can be brought to earth. My spirits lift considerably.

Dublin, Spring of 1995

Alone at the Point, a Van Morrison concert. The crowd is on its feet, dancing. "Did you get healed?" he asks, onstage, a dozen musicians backing him up. Maybe. I can feel the tectonic plates shifting: something is changing in my being. I am opening up.

Toronto, Winter of 1996

I have been to a meeting of Alcoholics Anonymous, but I haven't begun to wrestle with the God thing. Not even close. I am beginning to be concerned about my problematic drinking, but I am not going to beg God for help. The way I see it: just because I want it gone doesn't mean it's God's garbage day.

Besides, I'm not sure I even qualify as an alcoholic. Two seasoned members, Bob and Linda, have taken me for coffee before the meeting. They ask me: am I comfortable declaring myself an alcoholic? I tell them I'm still unsure—the word is so loaded. I see them exchange a glance before one of them says: "This is normal."

Am I an alcoholic? I think of Mum's behavior. Mine is nothing like hers: raging, anesthetized. *Alcoholic*: the word makes me squirm.

Still, there are times when I drink way too much. Not often. But there has been an incident. Onstage at *Maclean's* magazine's ninetieth-birthday party, June Callwood said she didn't have a chance of being named editor decades earlier because she had "the wrong genitalia." I had too much to drink, and repeated the line to someone I work for. Bad move. True, but still not smart.

Bob and Linda have taken me to a "closed" meeting, for alcoholics only. They prepped me: "When they go around the circle, saying 'My name is X, and I am an alcoholic,' just say 'pass.'"

Down some dark stairs, into a brightly lit church basement, where we sit like kindergarten children, in a circle. My heart pounds. "Pass," I say. The whole room stares. My head is swimming. I want to run.

At the break, an older man confronts me: "You didn't identify as an alcoholic." I freeze. "This is a closed meeting for alcoholics only. Do you belong here, or don't you?" I vow never to return.

Toronto, Winter of 2002

Carol Shields is dying of cancer. I am writing a short profile of her,
and we sit together in the sunlight. She looks unspeakably tiny
and frail, her head wrapped in a scarf. Over lunch, I ask her if she
believes in God. "No. Human goodness is the only thing I believe
in. To me, it seems astonishing that people are as good as they are:
that's the surprising thing. I do feel this sense of goodness is part of
our human conversation—the biggest part of it."

Winnipeg, Winter of 2008

At dinner, I tell Jake that I want to go to rehab—maybe I should
wait for a sign? If I see a rabbit in the backyard tonight, I'll take it as
God's will. He likes this idea. For the two of us, rabbits are totemic
creatures. He calls me "Rabbit." He's the jackrabbit.

I wash the dishes, and turn off the kitchen lights. I take a look out
the back window. No rabbit.

Jake lights the fire, I light the candles. Let me take one last peek,
I say, before I sit down. I peer out, and there, on the moonlit snow, it
sits: my rabbit. Call me primitive: for me, it's a sign. That, and Jake's
hand holding mine. These are the reasons I know I'll go.

The Bahamas, Spring of 2008

Fresh out of rehab, a trip to Kamalame Cay, where Jake proposed
two years ago. Walking to the great house for breakfast, we come
across a small egg: a baby killdeer, just emerging. My bird. Here is
my small bird of recovery. I pick up the fragile shell: I will keep this
by my bed as a reminder.

That night, I offer to get Jake a drink at the self-serve bar at the beachside villa where we're staying. I pour him a gin and tonic. Then, with considerable restraint, I pour myself a cranberry and soda. At the last minute, defiantly, I add a generous shot or two of vodka. How bad can it be?

All of a sudden, a man is at my elbow. "Hi, my name is Bill."

"I'm Ann."

"Can you pass me the cranberry juice?"

"Sure. Anything else?"

"No, I don't drink."

I look at him: tall, healthy, in his sixties. "Nor do I," I say.

"Really? I could have sworn I just saw you pour vodka in that glass."

I put down the drink. "I think I need to talk to you. I just got out of rehab."

Bill and I spend the next half hour discussing sobriety. He has been a member of AA for many years. So has his wife. A close call.

Two days later, lunchtime, eating alone. Jake has gone fishing for the day. I bring the New Yorker for company. At the next table, an enthusiastic crowd of women, sipping margaritas and white wine, laughing. They order a second round. Damn it, I think, I'm having a glass of wine. I start to rise, heading to the self-serve bar. "Hi, darlin'! I just thought I'd circle by to see how you're doing." It's Bill. "Can I sit down?"

I'm beginning to think there's something larger than myself. Is this possible?

Toronto, November 3, 2008

I cannot stop drinking: two weeks, three weeks, and then I relapse. I don't drink much—a glass or two—but it calls out to me. I cannot stay stopped. I am broken: mentally, spiritually. Down on my knees before breakfast, in tears. In my journal, Carol Kushner's words: "Our awareness of God begins where our self-sufficiency ends."

At a crowded noon meeting, I am at the back of the room, quietly weeping. A woman approaches me. "Do you need to talk to someone?"

She picks me up for the evening meeting, and delivers me home. Opening the door, I hear the phone ring. "Hi, darlin'. It's Bill. Just found your number in a knapsack from the Bahamas, and thought I'd give you a call to make sure you're doing all right."

Seven months later, and he calls on the day I got on my knees. What are the chances? And I am down on my knees again, with prayers of thanks. Am I just hedging my bets? Who knows. It's too early to tell.

Try to get sober and sooner or later, you're going to be confronting the big kahuna of sobriety—namely, Alcoholics Anonymous. Sooner or later, the issue of God is going to be part of the conversation, and you'll be wrestling with your notion of spirituality.

Is spirituality the antidote to alcoholism? *Spiritus con spiritum*: was Carl Jung correct? I know, in my bones, he was.

I love the story of Jung corresponding with Bill Wilson, the co-founder of AA. The year was 1961. Wilson wrote to commend Jung on his guidance to a fellow named Roland H., someone who had suffered as an alcoholic. Jung had counseled Roland H. on recovery, saying that the only hope was his becoming the "subject of a spiritual or religious conversion." Wilson closed his letter by telling Jung: "Please be certain that your place in the affection, and in the history, of our Fellowship is like no other."

In his response, Jung said that Roland H.'s "craving for alcohol was the equivalent, on a low level, of the spiritual thirst of our being for wholeness, expressed in medieval language: the union with God." He writes: "You see, 'alcohol' in Latin is *spiritus* and you use the same word for the highest religious experience as well as for the most depraving poison. The helpful formula, therefore, is *spiritus contra spiritum*."

As Jungian Marion Woodman puts it, Jung "recognized the confusion between physical and spiritual thirst."

Walk into any beginners' room of AA and you'll bump into God talk. Step Two? "Came to believe that a Power greater than ourselves could restore us to sanity." Step Three: "Made a decision to turn our will and our lives over to the care of God as we understood Him." (And yes, it's always a "him.") There's one "ultimate authority—a loving God as He may express Himself."

As Wilson wrote in 1961, the phrase "God as we understand Him" is "perhaps the most important expression to be found in our whole AA vocabulary. Within the compass of these five significant words there can be included every kind and degree of faith, together with the positive assurance that each of us may choose his own."

What is my faith? Why does spirituality work? Try describing air or water. This is a difficult one.

The best I can do is this: My faith sustains me. It feels like a poultice on my heart. I take the largest comfort in practicing Vipassana meditation, and I often pray to a creative force of goodness. I take solace in Woodman's words when she writes: "We live in a predominately Christian culture which has lost its living connection to the symbolism of wafer and wine. Lacking spiritual sustenance there is a genuine hunger and thirst. The archetypal structure behind the wafer and wine is slowly giving way to a new configuration, but we are in chaos during the transition. That chaos breeds loneliness, fear and alienation."

Amen.

And to battle that loneliness, fear, and alienation, so many of us are cobbling together our own microsystems of something that sustains us. If it's true that we are all in transition—and I believe we are—my own faith is still under construction. In my search for a deeper understanding, I turn to a variety of writers for guidance: the Jungians Woodman and Robert Johnson and Marie-Louise von Franz; the Buddhists Pema Chödrön, Jack Kornfield, and Thich Nhat Hanh, among many others: Annie Dillard, the poets. It's a wide-ranging search. As I said, it's a work in progress.

I cannot explain what happened to me on November 3, 2008, when I got down on my knees. But I know there was a huge internal shift. I was so broken by that point—emotionally and spiritually—that my soul just cried uncle. And for the next two years, my faith evolved—it's still evolving.

At first, I just prayed to stay sober, and to survive. I prayed for some answer to this simple question: why do I need relief from consciousness—or self-consciousness? Why was it a relief to be tipsy, or worse? Why did I always wake with a hangover untroubled, unknotted, and serene? Yes, it's true. And that's how it felt, almost each and every time, until midday. Then would come the slump of dismay. By six o'clock I could have a drink, and the process would begin again. By the time I quit, I was exhausted by the journey and all my prayers were about survival.

When I was in rehab, there was a moment near the end when the director came to speak to us as a group, about our drinking and our return to real life. He addressed us as you would small children who had committed a minor misdemeanor—like the time my sister and I ate the backs off all the Easter cupcakes Mum had been hiding in the cupboard. That sort of mistake—not the "you-fucked-up-your-entire-life" sort. The director advised us to make our beds each day, to join a group and keep our calendars full. Faith was not part of the discussion.

Actually, in rehab, I felt like one of those cupcakes: from the front, I looked whole; but from the back, I was gnawed apart. In our group, we would struggle to feign confidence. But alone? We were chastened, terrified souls, so uncertain of our futures that we wept like babies. All of us had made mistakes. All of us had suffered at the hands of others. Once in a while, someone would start to cry at mealtime and have to leave the table.

In rehab, I needed only marginal faith: that I was doing the right thing, making a 180-degree turn; hanging in, in the holding tank. I woke each morning, pressed play, and listened to Bach's "Sleepers, Awake." Did I pray? I know I did—even if, at times, it felt like I was just playacting. (I didn't, as Jake's sister Babe once said, believe in "scaffolding": a ladder up to heaven.) I meditated on a daily basis. And I sampled the full banquet of recovery group options: Rational Recovery, Alcoholics Anonymous—and Women for Sobriety, a group I led.

Life was raw—but not half as raw as it would be once I returned home, where I felt like a freshly peeled carrot, exposed and glistening. Once home, I found my energy started to play hide-and-seek on me, and depression dug in. I struggled with getting down on my knees. I struggled with a God that would allow so much pain and grief.

On those early dark days, one of my few comforts was reading. Wilfrid Sheed's recovery memoir, *In Love with Daylight*, was a favorite. Sheed, who suffered from depression as well, wrote, "Giving up booze felt at first like nothing so much as sitting in a great art gallery and watching the paintings being removed one by one until there was nothing left up there but bare white walls." He also said, "Booze is like an exit door painted on the wall for which alcoholics and other optimists manage to fall every time." That helped me stay sober. Like me, Sheed had a very tough early sobriety. For him, AA was little comfort—a place where "parrots" lectured to "sheep," and people talked "in bumper stickers."

I also loved E. B. White's writing, in which he speaks of the "flashy tail feathers of the bird courage." Once in a while, I would catch a glimpse of those flashy tail feathers, and vow to keep going. White gave me hope, and so too did Anne Lamott. I devoured her fabulous essay "Thirst," in which she gets sober and lets "a bunch of sober alcoholics teach me how to get sober, and stay sober. . . . It turned out that there were not going to be any loopholes. The people who seemed to find loopholes were showing signs of failure; for instance, they were shooting themselves in the head."

And so I too let a bunch of sober alcoholics teach me how to get sober, and stay sober. In doing so, I had to dig a lot deeper when it came to faith. I had to excavate the very foundation of my belief system, and rebuild it anew. On and off, the first eighteen months of sobriety were strenuous and difficult: my depression and anxiety just got worse. And then things turned hellish in the extreme: Jake broke up with me on the phone, one Monday morning, and I was stripped to my naked core. I was heartbroken with life, and solace was nowhere to be found.

From that day on, I had no choice but to rebuild. Within three weeks, I did another "hair-shirt holiday": I went into the woods and did a ten-day silent Vipassana retreat. One breath at a time, I began to reconstruct my life, one that looks totally unlike the one that came before. Day by day, month by month, I began to shape a life where I have a writing voice, where I am depression-free and happy and sober. Missing Jake still, daily. But fervently, deeply alive and thriving.

I did not do all of this alone. I had the help of several key friends—a group I call "the Three Horsemen of the Apocalypse"—and a superb doctor. Add to this months of effort and trust and, yes, faith. Faith that there would be better chapters ahead, faith in the goodness of people, faith in the goodness of my own bruised heart. Like Carol Shields, I believe that goodness is the biggest part of our human conversation.

And I have faith in a God of my understanding. I pray for a soft heart. My prayers seemed to be answered: what began as a supplicant process, down on my knees, has evolved into a dialogue.

I know that I have recovered my true self. That's the greatest gift of sobriety: the journey inward—endlessly challenging, rewarding, and profound. More often than not, I feel at peace in my own skin.

A major part of my evolution has taken place in reading and in conversation with others—an extended inquiry into what sustains them. For this book, I set out on a small quest: to speak to four people who had helped shape my own belief system. One of the sober alcoholics who helped me stay sober is Claire, a corporate lawyer in her early forties. I want to know, how did she wrestle with the God thing? She invites me to her colorful midtown Toronto apartment on a Friday night, a place where we can be private. Sitting in her living room, I see an easel with a half-finished canvas in the adjoining dining room, paints lined up across the table. Everywhere there are books and photographs, evidence of a life fully lived. In the past year Claire has hiked in Bali; in the past week, she has traveled to Mexico. But tonight, her full concentration is on the question at hand.

She settles back into the couch, collects herself, preparing her case. When she speaks, she speaks in paragraphs—perfectly formed. "When I got to the point where I was ready to ask for help, I was broken on many levels. Physically, I was sick from alcohol. Mentally, I was paranoid and my perspective was dramatically off. Emotionally? I lived in constant fear and anxiety and was deeply sad—it was like a soul sickness. My spirit was broken. I had lost my desire to live. I had no self-respect. At the end, I was trying to kill myself—that's what the drinking was: a slow death. I tried to commit suicide once, but I wasn't successful. But I could do it slowly.

"In terms of healing, the physical was the easiest part: I slept well, ate better, it just took time. The mental and emotional and spiritual are intertwined, and I'm still working on my recovery in those areas. I don't feel alone. I have a concept of a creative force—much bigger than me, that some people call God. That has given me great comfort because there are places that we go emotionally that no one else can come with us. In those moments, when I am afraid or my confidence is low, that's what I lean on. In the past, I leaned on alcohol. Therefore, when I hear Carl Jung's letter to Bill Wilson, that's what I think of: the concept of a loving creator has replaced alcohol as a solution.

"And it's not just the concept of God, but living life according to spiritual principles. I have a sense of purpose and a sense that I belong. Those feelings of not belonging still return—it's not like I got sober, found God, and all of a sudden had a full sense of purpose. I still have times when I am discouraged. In early recovery, I was full of anxiety every day. I had panic attacks. Now, things aren't so overwhelming."

And God? "I was raised a Catholic, and I had a very strong sense of a Christian God—which at no time did I reject. I just simply *forgot*: I lived my life." She smiles, and when she does, she looks like the French movie star Juliette Binoche. "As I started to drink more, I simply didn't develop that relationship. I didn't believe that God would come down and fix me, so I didn't look up." She grins at her own foolishness. "The fellowship was my higher power—all the other people who had recovered. It gave me hope, and that was enough to move forward. Within time, I started to thaw, and things got clearer. I started to see something at work in the rooms of recovery that I could not explain—and it was the start of a belief in a higher power.

"I don't know when I started believing in God. I just started praying down on my knees." She smiles and leans back. "It happened so quietly. Faith is fundamental to my recovery." She has argued her case, and I am won over.

One of the great advantages of being a writer is the ability to phone someone you deeply admire—a stranger—and ask if you can meet. In early sobriety, I found *A Woman's Way Through the Twelve Steps* deeply liberating. Stephanie Covington's generous, flexible attitude and open heart won me over. So I asked if I could visit her at her home just outside San Diego, to talk about spirituality and her own voyage of recovery.

Enter Covington's private perch above the ocean in Del Mar and you are transported into an elegant, rarefied world, with a profoundly Eastern influence. There is folk art, Buddhist art, and artifacts from around the world. More than one figure of the goddess of compassion—Quan Yin—welcomes you.

And of course, there is Covington. She greets me with a warm embrace, and there is deep connection in those piercing blue eyes. It is not the first time we have met. I spent some time with her at an addiction conference two summers ago when she gave an excellent presentation on trauma. But there is something essential about entering a person's private haven. I am unprepared for the beauty, the care with which each room has been shaped. This is a home that has been loved into being, and it radiates peace.

This is a treat: I know that three hours in Covington's company will be enlightening. She has an enormous perspective on all things related to women and addiction. When I arrive, the kettle is on the boil. Within ten minutes we are nestled on her couch, drinking delicate Thai blue tea, something she discovered in Paris with her daughter. When I ask her about the relationship between spirituality and recovery, her eyes are knowing, her voice soft and certain. "Addiction attempts to fill a spiritual void—and of course, it doesn't work. Finding spirituality is important for sustained recovery. At the core of recovery is the spiritual piece."

Covington, who has been sober for thirty-four years, is concerned that our culture is fostering an addicted society. "We are being taught to want more: our society supports compulsive behavior—overeating, overshopping. We had the Age of Repression, the Age of Anxiety, and now the Age of Compulsivity, where *some* is not enough and *more* isn't better. The spiritual part is necessary for balance, and balance can be a lifelong struggle."

What prompted her own search for balance, more than three decades ago? "The overwhelming majority of women in recovery are trauma survivors—but I am not one of them." When Covington faced her own alcohol issue, she was wearing another skin: wife of a Wall Street investment banker, working in the antiques and design business, playing tennis and bridge, the mother of a young son and daughter. Before she had graduated from college, she told two people she might be an alcoholic—"I could drink a lot, but I always got sick." In her early years as a parent, her drinking progressed. One night she had a fall and a blackout. "I got up the next morning and I had a gash across my face—and I didn't know how I got it." She called AA. She says: "Part of it was my eleven-year-old son's demonstration of what I had looked like, and it was humiliating; I was clutching at chairs as I tried to stay standing. But when I went to Alcoholics Anonymous, I didn't go to stop—I thought there was going to be a better way to drink—I went there to learn how to drink less!"

One month into her recovery, she "felt free. I felt lighter. *Euphoria* would be too strong, but the monkey was off my back. And I thought: how can I give this sense of possibility to one other woman?"

In recovery, Covington began a total shift in her life. Alcoholics

Anonymous advises that you make no changes for the first year. "I re-member saying: 'If I stay sober, I can look after my children and leave within a year.'" Leave she did: someone came back into her life and she returned to her home state of California. Ready to work in the addiction field, she didn't have enough sobriety to do so. What she did have was a master's in social work from Columbia University, so she focused on getting a Ph.D. in addiction and the psychology of women. Covington has been working in addictions ever since, becoming a leading special-ist in trauma and women's recovery. "The way I am today is not who I was," she says, with distinct modesty. "I believe every powerful life experience stays with us—good and bad—and that's why we say 're-covering,' not 'recovered.' My addiction was the best gift of my life: it's the thing that shook me and made me wake up—it got me conscious. I am very grateful. I think if I hadn't got into recovery, I'd be dead. I'd periodically think about suicide—but I had children."

One of her major achievements has been her contribution to the understanding of the role of trauma in women's addiction. Another major contribution is *A Woman's Way Through the Twelve Steps*, which has sold more than four hundred thousand copies worldwide and has been translated into several languages. "At the end of the eighties, things were written, saying the Twelve Steps weren't good for women because of the first step and its reference to powerlessness: 'We admit-ted we were powerless over alcohol—and that our lives had become unmanageable.' One young woman told me that the Big Book must be about two hundred years old. A lot of professionals thought AA was crap, a cult. What motivated me was working with clients who were having trouble going to meetings because of the language—and my own irritation with sexist and reductionist language."

She turned to thirteen women, including "four radical feminists who had used the Twelve Steps in a way that was not in conflict with their values." One was a professor emerita, a specialist in feminist ethics, who said: "The power of this program is in the spirit *beneath* the words." Their voices are woven throughout the book, along with Covington's very reassuring interpretation of recovery: "This is what it's about: integrating inner and outer, and thereby creating integrity."

The book has several references to Covington's "mother guilt." "I was a distant stranger to my kids because I was lost in an alcoholic fog," she writes, bravely. "When I became sober, it was painful to see how alienated we were from each other. I hadn't been available to them, and I hardly knew them. Children need attunement and a sense of attachment—and I didn't have the capacity. First, it's sad: but what are you going to do about it? This is where living amends come in. This means learning to be more present. I am a better mother today than I was when they were young. I have learned to focus on them and become attuned."

How do we heal? "We're all born with innate qualities and characteristics, and then things happen to undermine or cloud them. Addiction overshadows and impacts everything about the inner self—our thoughts, our beliefs, our values.

"Ultimately, recovery's about transformation—and I don't think it happens in isolation. I believe women recover in connection with other women. It's primarily about this connection, and the connection with a higher power. Many people will ask: why isn't it just about willpower? I believe there has to be an internal shift. It's about growth and expansion."

Is she religious? Was she brought up in a religious family? She chuckles: "Religion? I was raised a Protestant. I remember my mother saying: 'I am so glad we belong to the Methodist Church—such a wonderful sanctuary for a wedding.'" She pauses, and then says simply: "I don't think we do it alone."

The third person I speak to is the brilliant Gabor Maté, author of *In the Realm of Hungry Ghosts: Close Encounters with Addiction*. Lying in bed, one snowy Sunday morning, I am reading these words in the Vancouver-based addiction doctor's wonderful book: "The word God could have a religious meaning for many people. For many others, it means laying trust in the universal truths and higher values that reside at the spiritual core of human beings, but are feared and resisted by the grasping, anxious, past-conditioned ego."

Impulsively, I send him an email: could he talk to me in the next two weeks? Minutes later, the phone rings: "It's Gabor." I ask him about spirituality. What, he says, do I mean by spirituality? I am caught off guard. He says: "'Spiritual' is nothing more than liberation from the personal history. Underneath the personality, there's a deeper self. If you get back to that truth, that compassion, there is something essential. We're all part of something bigger—which the Twelve Steps refer to as the higher power. The Twelve Steps work because they speak to essential truths. People don't like to believe in a guy with a white beard in the sky—and they don't have to."

Why, I ask him, do we stop growing when we are traumatized? "Traumatized children shut down emotionally—and they also stop developing emotionally. Why? Because you need emotional vulnerability to grow. We are like crabs: we don't grow where our bodies are hardened. The greatest loss is not that there was pain. The greatest loss is that we lost a connection to our essence. That's our wound: the loss of connection to ourselves. When you recover, what do you recover? Yourself."

Finally, I speak to Karin, a woman who has thrived in Alcoholics Anonymous. Ask her why and she has a one-word answer: "connection." It reminds me of E. M. Forster's epigraph to *Howards End*: "Only connect!"

Karin is sitting in her living room, her dog by her side, fire blazing, her tidy frame tucked up on the couch. "This is a disease of percep-tion, and this is likely the first time in our adult lives that we have connection and unconditional support. If you look at the science of happiness, it's about being connected to others—AA is an antidote to the isolation of drinking." She looks troubled, and then refers to her own childhood history: "Of course, some of us had a connection and that connection abused us."

Why do people isolate when they drink? "Shame—and the fear that no one they know will let them drink the way they want to drink. Whenever our addiction took on a life, we stopped growing emotionally. Somewhere along the way, we veered off maturing, and we couldn't stand ourselves. When I look at my defects of character—self-pity, the urge to isolate—that's when they kicked in: when I started drinking at eight or nine.

"The program and steps give us tools to learn how to connect to people and how to live with other people. AA says: This isn't about willpower—there isn't an alcoholic alive who didn't try to use their willpower! You hand your power over to a higher power. Addiction is timeless, and this program is timeless. Just because you can tweet about it doesn't mean that the lessons about human bonding have changed. There are Buddhists and Muslims and Jews all sitting in the same room, feeling safe and accepted, learning to see our true selves and to love what we see. We connect to ourselves."

17.

Stigma

A CALL TO ACTION

It's the misunderstanding that kills.
—SUSAN CHEEVER

You could say to me, "Drink responsibly," and I'll say: "I'll try!"
—CRAIG FERGUSON

In our society, would you rather be known as an alcoholic or a person who suffers from depression? I have posed this question to dozens of women in the past three years. The answer? Not one woman chose alcoholic. To a person, they felt the stigma was too overwhelming. And for that reason, very few of the women I interviewed were willing to share their real names. Many considered it, but most chose to remain invisible. The rare ones—Beata Klimek, Vera Tarman, Annie Akavak, Marion Kane, Janet Christie, and, yes, my own brave mother—deserve my undying respect. They came on this journey with me, and it took true courage, real grit, to share their full names. These were the ones who decided to stare down stigma, and help others in the process.

The stigma is enormous. When I was preparing to write my

fourteen-part series on women and alcohol in the *Toronto Star*, I did so without once identifying as an alcoholic myself. Not that I didn't consider coming out in print. In my first meeting with my editor, I told her the truth over lunch. Her eyes widened, and then they lit up. There was no doubt: she wanted the story. To her everlasting credit, she helped me weigh the pros and cons. In the end, she summed it up this way: "When the series is over, will you need a job?" "Yes, of course," I said. I wavered. "Do you need to work, Ann?" "Definitely," I said. Her response was emphatic: "Then don't do it."

I loved her for that. I still do. She is both a perfect editor and a stellar human being. However, this wasn't the reason I kept my story to myself. The focus of my series was alcohol policy, and I felt the reader might doubt a writer who was herself abstinent. I chose to keep myself out of the story.

With this book, all that changed—but it wasn't an easy decision. There were many times when I would wake at 3 a.m. and think: "What sort of person wants to tell these stories on *themselves*? On their own mother?"

It took me six months before I could hand a book proposal to my agent. It took a lot of coaxing. But once I found my voice, I was swimming in the cool, clear water of truth.

For me, coming out is right. For so many years, I lived with the so-called secrecy of my mother's story. The sulfur fumes of smothering that reality poisoned our entire family. This is not my mother's fault. It's the legacy of every family that squelches the story of addiction, and so many do.

Which doesn't mean it's easy to say I'm an alcoholic. When I was in rehab, heading off to a variety of evening meetings, my Buddhist counselor Terry advised me to say: "Hello, my name is Ann. I am a *recovering* alcoholic." She believed our souls heard the words, and would flourish with encouragement.

And that's what I am today: a recovering alcoholic, whose depression is currently in remission. I am a professional woman, a mother, who has dealt with a few afflictions. It's a mouthful. It's a life-full.

Why is it easier for me to say I suffer from depression than that I am a recovering alcoholic? Actually, neither is *easy*. But only one could keep me from gainful employment. Only one invites universal judgment.

Why is this? Is it because I can take an antidepressant for depression, a pharmaceutical solution that others trust? Would it make it better if we medicalized the solution to alcoholism? Somehow, I don't think so. The taboo is still so large, the misunderstanding so deep.

When my series appeared in the newspaper in late 2011, the comments poured in, and there were two distinct schools of thought. Yes, there were laudatory emails, empathetic to the stories of addiction and recovery. My mailbox overflowed with moving personal accounts and positive messages: the series "should be required reading in every school in the country. . . . Don't stop writing." A woman who identified herself as T. J. Harrison wrote: "I hope that your series of articles spares future generations the anguish of ever having to try to recover from an intractable condition, and spurs thoughtful discussion of—and enlightened action on—a complex topic."

But while one reader wrote, "Alcoholism is a disease and it can break the strongest people," another wrote, "Addiction is not a disease. It's a personal lifestyle choice. . . ." "Stop calling it an illness," wrote another. "That's total bunk designed as a crutch for the weak . . . and this recovering rubbish is more bunk."

"Alcohol and drugs are the means for people who lack intestinal fortitude to face trauma," wrote another reader. "Stop glorifying addictive personalities and making excuses for lack of courage." Yet another wrote, "Alcoholism is not a disease. Cancer, diabetes: those are diseases. Alcoholism is self-inflicted. Grow up, take personal responsibility and learn to say NO."

More than one reader argued back: "So, you say alcoholism is not a disease? So it's the alcoholic's fault? We would love to be able to enjoy a drink or two responsibly, without any incident."

For decades, alcoholism has been what some call "a disputed ailment." Is it a disease? This is the perennial question—and it matters. So much of the stigma pertains to this debate: is alcoholism a sin of commission, or in the blood?

I asked Peter Thanos, a Canadian-born neuroscientist at the U.S. Department of Energy's Brookhaven National Laboratory on Long Island, New York. Thanos is blunt: "We have known for more than twenty years that alcoholism is a chronic, relapsing brain disease. Science supports this truth." Patrick Smith, CEO of Toronto's Renascent treatment center, is also unequivocal: "The jury is in. The Canadian Medical Association calls it a disease. The American Medical Association calls it a disease." "Our only shot is to see it as a disease because there's much less stigma," says the pragmatic Jean Kilbourne, herself in recovery.

But as Jungian analyst Jan Bauer writes in *Alcoholism and Women: The Background and the Psychology*, medical models do not accommodate the role of the psyche. "If, in fact, alcoholism is a kind of slow suicide, then it isn't surprising that collective attitudes are similar in both, especially in our Western society which, more than most, fears death. . . . [L]ike suicide, alcoholism expresses a refusal of life in the conventional, collective understanding of it."

Perhaps this explains why so many see addiction as a moral failure. Perhaps it's the volitional piece: you have to pour it down your own gullet. There was a time—a very long time—when I could not understand my mother's dependence. Why didn't she just *stop*? It would be decades before I was able to fully comprehend. By then, I was tortured by the same question, different pronoun: why couldn't *I* just stop? What led me to alcoholism, and not my siblings?

Thanos knows there is a genetic component to alcoholism. He has used brain imaging and behavioral studies with rats to understand human reward circuits. He has proven that certain brain receptors play a role in excessive drinking.

Here is how it works: alcohol, like all addictive drugs, increases the brain's production of the neurotransmitter dopamine, which sends a message of pleasure and reward. Over time, the brain responds to the stimulation of alcohol by decreasing certain dopamine receptors. These receptors—known as D2 receptors—are nerve cell proteins to which the dopamine must bind to send the pleasure signal. An alcoholic will experience a reward deficiency, and compensate by drinking more to try to recapture the pleasure. "Alcoholics will continue to drink to avoid the crash that comes with the low," says Thanos. "Alcoholics have lower D2 levels in their brains. If you have a lower D2 level, you are more vulnerable to the rewards of alcohol. And if you are genetically more vulnerable to the rewarding elements of alcohol, you are also more vulnerable to the atrophy of the brain from alcohol use."

If Thanos's research has proven there is a genetic component to alcoholism, it has also convinced him that there is more than one gene at play. He says: "Alcohol is a very dirty drug. The consensus is that ultimately, we will understand all the genes and then we will have to understand how they interact. We are just in the early stages of understanding the pieces of the puzzle of alcoholism."

Thanos and his peers may be at the early stages of understanding the puzzle, but the broader public lags far behind. Patrick Smith believes that social drinkers have a difficult time understanding the physiological realities of alcohol dependence because "it's not part of their lived experience." Still, as he says: "No one says: 'Just because I don't have diabetes, it doesn't exist.'"

Gabor Maté sees it a little differently. He believes the disease model is "valid, but insufficient." Says the Vancouver-based addiction doctor: "Why does someone develop the disease? It's trauma. Trauma isn't just an emotional event: it shapes brain circuitry. On brain scans, you can see changes shaped by trauma. And the medical profession doesn't get it. Brain changes are caused by early experiences. We are way behind the science in our understanding of addiction."

"For a woman, the stigma of addiction is still very bad," says Sheila Murphy, head of the women's programs at the renowned Hazelden treatment center. "And there's stigma for women who have gone to treatment. Which is a shame because women are taking time out of their lives to do some self-discovery, to understand on a deeper level how they want to live. It takes great courage to make changes in their lives."

Says Stephanie Covington, author of *Beyond Trauma*: "I find it interesting and sad: if I ask a group to think about addiction, I get them to think of a person who was or is an addict. For most of them, the person is not in recovery. It's all negatives: they ruined family holidays, cracked up the car, took money. You could never trust them. And that's what makes addiction different from heart disease or cancer—that's the attitudinal piece, the volitional piece. It's about choosing to drink or use. But no one *chooses* to be an addict."

"I'm very well-versed in stigma," says twenty-four-year-old Rebecca. "If you have an addiction, you're the scourge of humanity. A big process of my recovery was self-forgiveness. It's been a process of reframing, and not letting other people's views affect me—it's a huge obstacle to anybody getting sober."

Scout agrees. "I wish I could use my real name," she says. "But

I find anonymity important because people don't understand about alcoholics. If I were to tell my colleagues that I was one, they would think I wasn't up to the job. I tell no one—not even my family."

Lunchtime on a blustery winter day, and the meeting is packed to overflow. As it opens, there's a drip, drip, drip from the tap in the neighboring kitchen, but the room is still. The floor is now open for sharing. A middle-aged woman stands and announces: "I need to tell you that Anna, a member of my group, died last Friday of this disease. She was a beautiful woman, married, with two daughters. She decided she could do it on her own."

The room is very quiet. The next person to speak is a man with more than four decades of sobriety. "This disease kills, and it incarcerates." People bow their heads, absorbing the news.

Eventually, the meeting continues, but there's a question on many minds: Anna who? Days later, many will have figured it out from an obituary in the newspaper, featuring her picture. A beautiful dark-haired woman, whom many recognize. But for now, her anonymity has cloaked her identity.

The underground railroad that is Alcoholics Anonymous thrives on the promise of many things, not the least of which is anonymity. What this means, literally, is that individuals must keep quiet not about their sobriety, but about their membership in AA, or any other twelve-step program for that matter. For more than seven decades, anonymity has been key, ensuring a safe haven for sharing and recovery.

In 1935, when AA was founded in Akron, Ohio, anonymity made sense. But in the era of Facebook and Twitter, there are those who argue that anonymity is a dated concept. Is it time for AA to drop the second *A*? Would that address some of the stigma around alcoholism?

Some say yes—most famously author Susan Cheever. She has written not only about her own drinking and that of her father, writer John Cheever, but also a biography of Bill Wilson. In 2010, she wrote a controversial column in the online New York–based magazine the *Fix*: "We are in the midst of a public health crisis when it comes to understanding and treating addiction. AA's principle of anonymity may only be contributing to general confusion and prejudice. When it comes to alcoholism and AA, the problem is very public, but the solution is still veiled in secrecy."

If alcoholism is a disease, does anonymity promote a sense of shame that is outdated? Cheever obviously thinks so, especially as it relates to AA's Tradition 11: "We need always maintain personal anonymity at the level of press, radio and films." Says Cheever: "It seems to me that AA is the best treatment we have for alcoholism—not that it is perfect. But people do not understand what alcoholism is or that there can be recovery. And there are those who are dying because of this. Twenty-five percent of all our hospital admissions are related to alcohol."

Is she right? Long term, is AA the best treatment for alcoholism? Yes, says Patrick Smith. "In recent years, the penny has dropped," he says. "Post-treatment, those who remain involved in a mutual twelve-step program like AA dramatically improve their chances of remaining clean and sober."

What about harm reduction? Johanna O'Flaherty, vice president of treatment services at the Betty Ford Center, minces no words: "Harm reduction is bullshit. Addiction is a brain disease—if we cross into addiction, there is no going back." Vera Tarman, medical director at Renascent, agrees: "I see myself as an important dinosaur, a voice advocating abstinence. Although I want to give my respect to harm reduction, it's a passion of mine to promote the concept of abstinence as a positive thing—freeing, not restrictive. In the professional milieu, it's looked down upon as old-fashioned, too religious. We can make it new again. We need to make it new again."

Which circles back to the anonymity question. "Anonymity protects," says Cheever. "But it also hides." She draws comparisons to the gay world and the act of coming out. So does Maer Roshan, editor and founder of the *Fix*—which is focused on the world of recovery. "I think that the recovery world is where the gay world was in the 1990s," says Roshan. "Blacked-out windows and bars, a secret world." He believes there are good reasons for the anonymity rule: "If people relapse, and they are known, it destroys the notion that AA brings success. As well, you are supposed to enter the program as equals—as 'Peter,' not 'Peter Rockefeller.' No one has the right to expose another. But I do think it's a gray area when they say you're not allowed to speak of your own membership."

Covington has been in recovery for more than three decades. She thinks the only people who should break their anonymity are those like herself: those who have been sober long enough to be stable, not lose jobs or relationships. "I do think recovery needs a face and a voice," she says. "Maybe it behooves those of us who can't be hurt to be more vocal. It's too bad there aren't ways for people to understand the value of the program."

What Cheever raises is something many authors have had to wrestle with. In the world of "quit lit," many have broken their anonymity, most notably the late Caroline Knapp in *Drinking: A Love Story*. Others have walked a fine line, talking about meetings without naming the groups: Mary Karr in *Lit*, Susan Juby in *Nice Recovery*. Says Juby, who is more than two decades sober, "Anybody who knows anything about anything can presume. But I understand the sentiment around anonymity. It's not just about being private concerning personal issues, or protecting an individual. The tradition is also based on humility. I like the idea that recovery moves through example at a community level. There's a nobility to that. Still, the hostility to meetings always surprises me: the idea that it's a cult has lots of sway."

"It may be a cult," says Karin. "But if so, it's one I want to belong to. If it weren't anonymous, I wouldn't be here." The woman to her right nods. "There's a pecking order when it comes to stigma. If you're male and drunk, you're a good old boy. If you're female and drunk, you're not going to live it down."

Is AA too shrouded in mystery? As a journalist, Roshan believes so. "I think AA is a very under-covered subject. It's an institution unlike any other. It should be reported on just as the Catholic Church is, and the presidency. Who's being served by this current situation?"

Perhaps the individual members, says Pat Taylor. Taylor is executive director of Faces & Voices of Recovery, an advocacy group based in Washington, D.C. For a decade her group has helped tens of thousands of individuals in recovery, as well as their friends, families, and "allies," tell their stories in a positive manner through what she calls "recovery messaging." She agrees with Cheever on one point. "One of the greatest problems is that the general public and policy makers don't understand that people *can* and *do* get well," says Taylor. "They're moms and dads, pay their taxes, and they vote. Not that long ago, women didn't speak about breast cancer. Bringing recovery into the public health arena is critical." But never does this mean breaking anonymity, she says. "There are many pathways to recovery, and we support and embrace them all. We support whatever any mutual support group advocates."

William C. Moyers, one of the founders of Faces & Voices of Recovery, agrees with Taylor—and yet he revealed his membership in AA in his book *Broken*. "If I was going to explain the gritty details of the spiral of my addiction, I needed to give the nitty-gritty details of getting sober," he says. "Recovery is a mind, body, and soul experience. It's not magic. I owed it to my readers and their families to talk about the Twelve Steps and AA. Quitting is not about stopping drinking. I stopped a million times. It's about staying stopped, and it's hard to stay stopped."

Still, Moyers believes "it is not necessary for most to break their anonymity to get the word out about AA. You can be a voice of recovery without breaking your anonymity. AA should stick to what it's always done—which is people helping other people."

Cheever envisions a different world. "What if it was widely reported that a significant percentage of U.S. senators are in AA, or that there are AA meetings in the West Wing of the White House?" she writes. "What if hundreds of the movers and shakers in recovery—doctors and lawyers and airline pilots, the Fortune 500 businessmen and ministers—stood up and were counted as members of AA? It would go a long way toward clearing away the misunderstanding that still surrounds us."

The world is changing, and quickly. A series of online groups have been springing up around the world, offering support to those who want to stay sober—or give sobriety a whirl: Soberistas out of London, Booze-Free Brigade out of Los Angeles. One of the more interesting initiatives is called "Hello Sunday Morning." Three years ago, a young Australian named Chris Raine started this online group, throwing down the gauntlet to others: take a three-month break from drinking and see what happens to your life. Raine, a handsome dude of a guy who started in the advertising sector, still drinks—but each year, he takes his requisite break. He says: "Our vision is to have one hundred thousand members by 2015. The idea is, you don't need alcohol to be confident; you don't need alcohol to have fun; you don't need alcohol to be yourself. The majority of HSM members change their consumption long-term, learn how to re-form an identity to be the person who doesn't drink the most in their circle."

What Raine is helping to do is break the stigma around binge-drinking problems, opening up the dialogue. I hope the stigma will disappear in my lifetime. This year in Canada, I cochaired the first National Roundtable on Girls, Women, and Alcohol, a new initiative I helped found. Its aim is to open a public dialogue on the issues related to problematic drinking. We live in an alcogenic culture. We have normalized risky drinking. It should surprise no one that some of us fall through the cracks. We need to question: what are we doing to contribute to this reality, and what can we do to change it?

In the past decade, mental health has had significant leadership related to anti-stigma campaigns. I believe that addiction needs a similar response. But so far, most voices are silent. I hope to help change that. The gap between what we know about addiction and our perceptions of it: an embarrassment. People overcome addiction. They need to speak up, and they need to be heard. With a convergence of voices, so much could be won.

Becoming Whole

IN WHICH I RECOVER MY SELF

You will lose someone you can't live without, and your heart will
be badly broken, and the bad news is that you never completely
get over the loss of your beloved. But this is also the good news.
They live forever in your broken heart that doesn't seal back up.
And you come through. It's like having a broken leg that never
heals perfectly—that still hurts when the weather gets cold, but
you learn to dance with the limp.

—ANNE LAMOTT

Toronto, February 2013

I am not sure how men heal from addiction. For me, there has been
one central fairy tale that has been totemic in my journey: "The
Handless Maiden" by the Brothers Grimm, as retold by Jungians
Robert Johnson and Marie-Louise von Franz. It goes like this:

Once upon a time, there was a poor miller who lived with his wife
and his young daughter. As happens in fairy tales, the devil appears
in the guise of an old man, and offers to make the mill run faster, in
exchange for what stands behind the mill. Easy, thinks the miller: it's
my apple tree. He agrees to the deal. Immediately, the mill becomes
profitable. In short order, the miller becomes a wealthy man.

Three years to the day, when the devil reappears to claim his part of the bargain, the miller goes to fetch his axe. To his horror, he learns that his daughter had been standing behind the mill, sweeping the yard. The devil tries to take her, but her goodness manages to ward him off: she makes a circle around herself with chalk so he cannot approach her. Enraged, he commands the poor miller to chop off her hands. (In his wisdom, Johnson points out that the alcoholic "makes one of the worst devil's bargains," trading suffering for oblivion—or, in Jungian terms, the sacrifice of the feminine feeling.)

Handless, the daughter is exceedingly lonely, wandering every day in the woods by herself. By chance, she makes her way to the king's garden, where there is a prized pear tree. She reaches for one of the beautiful pears, and manages to eat it, an act witnessed by the king's gardener. The next day, he brings the king to see the beautiful handless maiden. Immediately, the king falls in love. He marries her, and has his magicians make her a set of silver hands.

And as we all know, silver-handedness is a bad bargain. "No other loneliness is as deep as silver-handedness," writes Johnson—and indeed, this is true. Anyone who has been dependent on alcohol knows this well.

In time, the queen gives birth to a baby boy. There are many servants to help with the care of her young son, but the queen is bereft. She wants to manage her baby by herself. One day, she starts crying and cannot stop. She takes her baby to the woods.

Writes Johnson, "As soon as the queen has bathed herself in the restorative bath of tears and gathered a reserve of energy, a most wonderful thing happens. The miracle begins as an emergency—as so many wonderful things do—when her baby falls into a stream and will drown if not rescued immediately."

The queen plunges her useless stumps into the water to rescue her child, and her own hands are fully restored. As Johnson writes: "The queen has instinctively understood that aloneness is better than false relationship—even if it be of sterling silver—and she takes refuge in the greatest of all feminine healers—solitude."

And what happens? Writes von Franz of the woman who has reclaimed her hands: "She will have full consciousness of what she is doing and is therefore rewarded for her suffering, which is what Jung means when he writes that 'a part of life is lost, but the meaning is saved.'"

Once I gave up the silver-handedness of my relationship with alcohol—the unrewarding habit of trading suffering for oblivion—I had to learn to fend for myself in the woods of life. New sobriety is a challenging experience if ever there was one: your first Christmas, your first New Year's, your first wedding or funeral. I have never felt more naked, exposed to my feelings, raw.

I remember a particular cocktail hour at the houseboat, when I literally did not know how to breathe through the pre-dinner period. I sat on the little driftwood bench, feeling fidgety and vulnerable. I remember a fancy celebratory dinner in Chicago, with Nicholas: when the waiter uncorked a good bottle of red wine a foot from my face, at the table snug beside ours, Nicholas touched his foot to mine, in solidarity. Half an hour later, when the wine kept flowing, he offered to walk me around the block. I accepted. I will never forget the tenderness of my son, linking an arm through mine as we left the restaurant and headed out into the night air.

Most of all, I remember the depression of early sobriety: feelings raw, and the constant tears. I mourned my relationship with alcohol. At heart, learning to live sober is a solitary experience.

Today, my so-called Celtic Blood Disorder is solidly in my rearview mirror. I hang out in church basements, with joy. Each time I take my seat at a meeting, I find myself smiling. I never take it for granted, this daily reprieve from addiction.

Not that there haven't been times when I would have killed for a drink. I was only eighteen months sober when Jake broke up with me, severing our deep fourteen-year connection. That first night, I would have given my right arm for a scotch. I had tea. There are nights when that wound can still ache.

Seven months later, at my father's wake, I watched wistfully as others soothed their grief with Pinot Grigio, my brand. Again, I would have killed for a drink, a little social novocaine to numb the pain. I drank strong coffee instead.

There were a lot of tears on both occasions. After losing Jake, I was angry at my sobriety for a long, long time. For a while, I found it tough to sit on those hard chairs under bright lights, and be grateful: I had lost so much. Before my world filled back up again, it became very, very empty. It often felt like holding sand in a squeezed hand: everything was rushing out. It took a long time for me to learn how to release my grip, to loosen up with life.

For some, early sobriety is a pink cloud. Not for me. In my case, I rode the dragon of depression and anxiety with difficulty. It took me more than two years to find my equilibrium, and a full three and a half to soar.

In the beginning, you think it's all about giving up alcohol. I remember counting out the days with pride, thinking: there, I've got it nailed! One hundred, two hundred, three hundred, a full year.

But slowly, you realize that abstinence is only the first requisite for growth. This is how you get ready for the real work: the emotional, mental, and spiritual push-ups essential to gaining some true perspective, some maturity. They say you stop maturing when you drink, and I believe it's true: I had a lot of catching up to do.

Two years ago, I was on the phone with my son in Brooklyn, lamenting all that I had lost when I gave up drinking. He was very quiet. Then he told me to get a piece of paper. "Mum, draw a line down the middle. I'll dictate this to you." Dutifully, I did as he said. How odd, I thought: this boy-man being my tutor.

"On one side, Mum, write 'Losses.' Okay, put Jake's name there. You lost the man you loved. And yes, he was a really, really great guy—and then he wasn't, Mum. Not to you."

"Now, on the other side, write 'Gains.' Write this, Mum. You got your son back." My heart is in my mouth. "Mum, I wasn't really even speaking to you. Our relationship was really strained."

"Is this true?"

"You know it's true. We didn't even speak for four months. Don't you remember? So write this: You supported me and my wish to go to art school. You have been a fabulous mother." Now I am silent. "You got your sister back, Mum. You got your relationship with your mother back. You got your friends back. Name them, Mum. Gillian. Keep going, Mum." I scribble. The list is growing. "You got your writing back." What else? "Are you writing?" "Yes, I'm writing." "Put it all down, Mum."

He continues. I run out of paper. "So, you lost a guy, Mum. Have a look at the other side."

When you emerge from an addiction, you actually get to choose the parts of yourself that you will keep, and those you will have to lose if you are to stay sane and sober. This is a full rebirth, in the truest sense of the word—painful and complete. In AA, they say it takes five years to get your marbles back. But it's more than that: it takes several years to shape a new self.

Today I am a much calmer version of my former being, steady and humbled by all that I have witnessed and weathered. Things have evolved in ways I could never have imagined. I got my marbles back. Most days I know what to do with them.

Sometimes the universe sends you gifts. I have had several of these in sobriety—strange experiences that made me believe there was something larger at work.

Jake and I used to play a foolish lovers' game: if we gave each other permission to run away with anyone in the world, who would we choose? Jake's choice always changed, often depending on what movie we had just seen: once, it was Scarlett Johansson; another time, after seeing Woody Allen's *Vicky Cristina Barcelona*, it was Penélope Cruz. Me? It never changed. It was always the sad-eyed, devilishly handsome Irish actor Gabriel Byrne.

That summer of the breakup, I flew to New York for a Thursday evening. Nicholas had a photograph in a show in a gallery in Chelsea, and I didn't want to miss my son's first opening. Afterward, he and his girlfriend and I headed out for something to eat. They stood on the street, while I checked out possible restaurants. I walked into one and emerged with a grin: "You'll never guess who's sitting at the bar. Gabriel Byrne." "Go introduce yourself," said Nicholas.

And so I did. I told him the story. "My sweetheart just broke up with me, and we always had a deal: if I ever ran into you, I was allowed to run away." Gabriel Byrne fixed me with the saddest eyes I had ever seen. "You have a very tough time ahead," he said. "Have a seat." This is not how I expected it to go. "What are you drinking?" He had both San Pellegrino and an espresso in front of him. "San Pellegrino, with lime." "Tell me what happened." I gave him the short version, and he listened carefully. He was quiet for a minute. "I want you to get John O'Donohue's *To Bless the Space Between Us* and read 'For the Breakup of a Relationship.' Copy it on a piece of paper, put it in your pocket, and take it out and read it every day for the next year. It will help. John O'Donohue, Irish poet." "I have it," I said. "By my bedside."

He ordered himself another espresso. We sat together talking for what seemed like a long time. He told me of his shoot that week, of a thirteen-year-old boy who thought his heart was broken. I told him how Nicholas had had an opening in Chelsea, how his girlfriend had been mugged on a Brooklyn street the night before. When Nicholas and his girlfriend appeared, she with a badly bruised face, Byrne said: "Well, you're quite a threesome. You almost got your nose broken, you got your heart broken, and you are a star."

With that, he leaned down and whispered something to me, spoke to the bartender, then slipped away into the night. Hours later, when I went to pay, our bill had been covered.

Two months later, I walked into a retirement party at the Drake, a hip Toronto hotel, listened to the speeches, and rose to get my coat. An elderly gentleman approached me, someone I had never seen before, and have never seen since. "Please don't worry—it's all going to turn out fine," he said to me. "Do I know you?" I asked. "No, but you have the most broken heart I have ever seen. I want you to know this: It's going to be all right. He was your soul mate. Give it time. Keep busy. I promise, it will turn out well." "How do you know," I said, tears welling up. "I can see it," he said. "I have a gift—I always have. May I get you a drink?" "I don't drink," I said. "Neither do I," said the man. "I'm allergic. I want you to know you have a lot to do, a lot of people to help. It will be fine." "How do you know?" "Don't worry," he said. "You have a beautiful heart." "Who are you?" I asked. "It doesn't matter. Love will find you again, and when it does, he will be a much older man." And then he left. I got my coat and circled back to him. "Why are you telling me all this?" "Don't worry," he said. "Just remember: you don't always have to be right. That's your one flaw. But

you have a beautiful heart, and you will have more love. Remember:
he will be older, much older. This will take time. In the meantime,
you have a lot to do." I left the party, feeling peaceful. I call this man
my Clarence, after the angel from *It's a Wonderful Life*. Clarence—
whoever he was—soothed my broken heart.

There was a summer when Jake taught me to befriend the female
snake who made her home at the houseboat: how to hold her and
not be afraid. This went on for days, my learning to pick her up. One
morning I found her skin on the houseboat deck. I took it home to
Toronto: a little totem for my writing desk. Today, remnants of it sit
there still; a perfect emblem for the shedding I have done in recent
years. It sits beside a rock Jake gave me in the shape of a heart, and another little one my agent gave me when I started this book, engraved
with the word *hope*. Hope is what we gain when we give up drinking.
Possibility.

Am I a "grateful alcoholic," as they say in the rooms? Most days, I
am. For several years now, I have begun each morning with a gratitude
list, one that lists my challenges as well as my blessings—especially
my broken heart, which reminds me how much I loved and was loved.
There is some strange alchemy associated with gratitude. Somewhere
along the way of doing these lists, I fell in love with my life again.

Most of all, I have learned that the addict's lie is just that: an un-
truth. It goes like this: "I will always feel this way—therefore, I might
as well drink." And as long as you keep drinking, that lie keeps you
stuck. The world does not improve: it's a self-fulfilling promise. Stop
drinking, and there's no telling what will happen. Stop drinking, and
you can begin the process of loving yourself back into being.

The Jungian Fraser Boa once said, "You are standing on the shaky
sands of doubt. Stand on the firm ground of not knowing." That's

where I am today: standing on the firm ground of not knowing how the rest of my life will evolve. I may have lost the biggest love of my life, but I have regained my first love: writing. I have found my voice again. All things are possible. And without alcohol in my life, there are so many more chapters to come.

ACKNOWLEDGMENTS

This book was first conceived several years ago on a hot August day, when Jake MacDonald hammered together several fine pieces of weathered board and created a desk for me, one which I situated on an outdoor deck above a mass of pink granite, on the edge of the Winnipeg River. Beside me, rosy cosmos nodded in the sun, tipping their heads toward the shoreline where pink rock meets dark green water. It was there that I began wrestling with this book: I was fresh out of rehab and hungry to find a narrative in my drinking experience. It was there that Jake told me to "drag all of the material out of the lake, onto the shore."

And drag I did, all the rich animal and mineral matter from my drinking days, what came before and after. Those first essential notes emerged in Minaki, as so much of my writing always did. Those few pages formed the nucleus of this book. But it really took shape by the edge of Jack Lake, at another retreat in the Canadian Shield, this one owned by my generous friend and former husband, Will Johnston. I spent many months in his cabin, alone with my dog, enjoying the silence in which to hear my own voice. I am deeply grateful for this gift of solitude, the company of white pines and a green canoe.

What began by the water was well nurtured, first by the

extraordinary generosity of the Atkinson Foundation. In 2011, my Atkinson Fellowship in Public Policy culminated in a fourteen-part series on women and alcohol in the *Toronto Star*. I am deeply indebted to that newspaper, the Honderich family, the Atkinson Foundation, and my editor, the talented Alison Uncles.

Still, this book would not exist without the enormous persistence, faith, and instinct of two extraordinary women: the sublime Hilary McMahon, my beloved agent, of Westwood Creative Artists; and the remarkable Karen Rinaldi, my wise and masterful lead editor in New York.

For months Hilary pestered me to proceed with my proposal. My inbox is full of messages entitled "Nagging" and "Checking in." And check in she did, over and over, until I found the guts to go forward. Simply put: without Hilary, there would be no book.

Karen Rinaldi believed in this book from the very beginning and has been a constant champion throughout, helping to shape what is before you today. She led the fight for this project, along with two other gifted women, both of whom contributed enormously: Iris Tupholme in Toronto and Clare Reihill in London. Those in the office of HarperCollins in Sydney rounded out the team. Indeed, Harper-Collins wrapped its arms around me and welcomed me from posts around the world, giving me a great sense of purpose and belonging. I am hugely indebted to my publisher for its generosity and support.

This book is grounded in a depth of knowledge that would have been impossible without my Atkinson experience. For a superb education, I am enormously grateful to the following: Jürgen Rehm and Norman Geisbrecht of the Centre for Addiction and Mental Health; Tim Stockwell of the Centre for Addictions Research of British Columbia; Michel Perron and Gerald Thomas of the Canadian Centre on Substance Abuse; Robin Room of Australia's Centre for Alcohol Policy Research, Turning Point Alcohol and Drug Centre; David

Jernigan of the Center on Alcohol Marketing and Youth; Sharon Wilsnack of GENACIS; Denise De Pape of the B.C. Ministry of Health; Nancy Poole of the B.C. Centre of Excellence for Women's Health; and Nancy Bradley of the Jean Tweed Centre.

For sharing their own stories, I owe a huge debt of gratitude to the many brave women in this book. A rare handful joined me on the journey, going fully public. To these women, I owe my undying admiration: Beata Klimek, Annie Akavak, Janet Christie, Marion Kane, Vera Tarman, and, above all, my own mother, Maxie Dowsett. I can only hope that their courage will meet with the deepest respect.

I wrote this book with the support of many friends, not the least of whom are Gillian MacKay Graham and Ron Graham. Both shared their home and their meditation "hut" on many occasions, celebrated my successes with enthusiasm, and read the book in draft form. For all this and more, I am deeply grateful.

Judith Timson was my constant companion throughout, feeding me stories, love, and daily support. I treasure her beyond measure, as I do Marci McDonald, Jane Stewart, Barbara Kofman, Tecca Crosby, Heather Arnold, and Nancy Hamm. Gail Heney, Kathleen and Caitlin Glynn-Morris, Elena Soni, Sarah Milroy, Pearse Murray, Linda Loving, and Caitlin MacDonald all played a pivotal role in my sobriety story. For their support, I am truly grateful. My book club sustained me with love throughout, as did my recovery group. To these friends, and many more, my never-ending thanks.

For his profound love and nurturing, I am deeply grateful to Jake MacDonald. For fourteen years, we were joined at the hip, hand, and heart. He taught me how to cast my rod and fish in the deep waters. His impact on me is indelible.

For their sturdy faith in me, I want to thank the late Carol Shields who, along with Marjorie Anderson, included my personal writing in *Dropped Threads 2: More of What We Aren't Told*; and my many gifted editors and colleagues at *Maclean's*, my professional home of more than twenty-five years.

Finally, I want to thank my bighearted family for their unflagging support and love. To my sister, Cate, for her constancy, keen eye, and care; and to my brother, John, for his wicked humor and spirit. Both read the book as it unfolded, and I am deeply grateful for their insight and perspective. Together, we are united by so much more than blood.

To my precious father, who was not here to witness the healing that evolved, I am grateful for his love, especially of our mother. To my darling mother, who has won my profound devotion and respect: for consenting to let me tell her story, from the very beginning. For that history, I am deeply grateful; for her bravery, I remain nonplussed. She is one of my closest companions, loyal and true.

To Will, who offered shelter and companionship at key junctures. If all failed marriages ended in such deep friendship, the world would be a better place.

And finally, I want to express my everlasting thanks to my son, Nicholas, who cheered me through the toughest spots and brought his wisdom to bear on this book. As a creative thinker, he inspires me. As a son, he delights me. His sense of adventure, his ample heart, and his deep love make my life a profound pleasure.

With all of this, and so much more, I am deeply blessed.

NOTES

ix   "Our excesses are the best clue": Adam Phillips, "The Insatiable Creatures," *Guardian*, August 8, 2009.

xi   Charles Bukowski, "The Laughing Heart," *Betting on the Muse* (Harper-Collins, 1996).

PROLOGUE

1   Hang out in the brightly lit rooms of AA: This essay first appeared in the *Toronto Star* in November 2011. It is reprinted with the generous permission of both the newspaper and the Atkinson Foundation.

CHAPTER 1: THE MONKEY DIARY

7   "To be rooted": Simone Weil, *The Need for Roots* (Routledge Classics, 1952).

8   Donated to McGill University: "Seagram Building Reborn as Martlet House," *McGill Reporter*, March 24, 2004; http://www.lonelyplanet.com/canada/montreal/sights/architectural-cultural/seagram-house.

8   in charge of development, alumni, and university relations: "Telling McGill's Story," *McGill Reporter*, April 13, 2006.

20   "I said to my soul, be still": T. S. Eliot, *Four Quartets* (Harcourt, 1943).

CHAPTER 2: OUT OF AFRICA

23   "One always learns one's mystery": Robertson Davies, *Fifth Business* (Macmillan of Canada, 1970).

24   Year after sunburned year: Much of this cottage memoir first appeared in *Maclean's* magazine, August 20, 2001. Reprinted by permission from *Maclean's* magazine, Rogers Publishing Limited.

26   Hey Mabel, Black Label: http://www.youtube.com/watch?v=cnhQxUKqrMY.

36   alcoholism the black lung disease of writers: Tom McGuane's quote appeared in Jim Harrison's *Off to the Side: A Memoir* (Atlantic Monthly Press, 2002).

CHAPTER 3: YOU'VE COME THE WRONG WAY, BABY

39   Alcohol abuse is rising in much of the developed world: Interviews with Jürgen Rehm, May 2010 and May 2011.

40   the top 20 percent of the heaviest drinkers consume roughly three-quarters of all alcohol sold: Interviews with Gerald Thomas, May 2010 and May 2013.

40   According to a recent CDC: Interview with Robert Brewer, February 2013; *CDC Vital Signs*, January 2013.

41   In Britain, Prime Minister David Cameron: BBC, February 15, 2012.

41   Dame Sally Davies: *Telegraph*, November 21, 2012.

41   British girls were cited as the biggest teenage drinkers: *Guardian*, April 22, 2011.

41   "In the thirty years I have been a liver specialist": Interview with Sir Ian Gilmore, February 2013.

42   multibillion-dollar international industry: Interview with Robin Room, December 2010.

42   German researchers found: *Alcoholism: Clinical & Experimental Research* 37, issue 1, (January 2013): 156–63.

42   On average, both men and women died roughly twenty years earlier: Interview with Jürgen Rehm, February 2013.

42   "It is just like Virginia Slims": Interview with David Jernigan, February 2013.

44   "Women's economic empowerment": *Economist*, December 30, 2009.

44   Sheryl Sandberg: *Lean In: Women, Work, and the Will to Lead* (Alfred A. Knopf, 2013).

44   Anne-Marie Slaughter: "Why Women Still Can't Have It All," *Atlantic Monthly*, July/August 2012.

45   Meanwhile, in the United States, two-thirds of married male senior managers have children: http://www.ted.com/talks/sheryl_sandberg_why_we_have_too_few_women_leaders.html.

45   "Sexism is invisible, but it's real": Interview with Daisy Kling, May 2013.

46   "Leadership Ambition Gap": Sheryl Sandberg, *Lean In: Women, Work, and the Will to Lead* (Alfred A. Knopf, 2013).

46   review of *Lean In*: *New York Times*, "Yes, You Can," March 7, 2013.

46   "Having it all, at least for me": Anne-Marie Slaughter, "Why Women Still Can't Have It All," Atlantic Monthly, July/August 2012.

49   women are less happy today than their predecessors: http://isites.harvard.edu/fs/docs/icb.topic457678.files//WomensHappiness.pdf.

49   the workweek of the typical middle-income American family increased: Joan C. Williams and Heather Boushey, "The Three Faces of Work-Family Conflict," Center for American Progress, January 25, 2010.

49   According to a 2011 study by the Center for Work and Family at Boston College: Referenced in "Why Gender Equality Stalled" by Stephanie Coontz, *New York Times,* February 16, 2013.

50   "global epidemic": Interview with Sharon Wilsnack, November 2010.

50   In 2011, Katherine Keyes: Interviews with Katherine Keyes, November 2011 and June 2012.

51   "Young professional women drink a lot more than women in manual and routine jobs": Interview with Katherine Brown, February 2013.

52   The largest health benefit comes from one drink every two days: Interview with Tim Stockwell, December 2011.

52   challenged the broadly accepted assumption that a daily glass of red wine: *Addiction,* http://onlinelibrary.wiley.com/doi/10.1111/j.1360-0443.2012.03780.x /abstract.

52   "While a cardioprotective association between alcohol use and ischemic heart disease exists": Interview with Jürgen Rehm, February 2013.

52   "But hormonally, metabolically, men and women are different": Interview with Joseph Lee, May 2011.

52   Women's vulnerabilities start with the simple fact that, on average, they have more body fat: http://alcoholism.about.com/cs/alerts/l/blnaa35.htm.

53   GENACIS: http://www.genacis.org/.

53   the strongest predictor of late-onset drinking is childhood sexual abuse: Interview with Sharon Wilsnack, November 2010.

53   consume four drinks and you will leave yourself vulnerable to compromising your spatial working memory: Interview with Lindsay Squeglia, November 2011.

54   "Are the girls trying to keep up with the boys?": Interview with Edith Sullivan, November 2011.

54   "It is *the* issue affecting girls' health": Interview with Nancy Poole, February 2013.

55   "Alcohol is not a women's issue": Interview with Gloria Steinem, June 2013.

CHAPTER 4: THE FUTURE IS PINK

57   Ingredients in Bitch Fuel: http://www.rulloffs.com/pages/m_drink.html.

58   My favorite drinking memoir: Caroline Knapp, *Drinking: A Love Story* (Dial, 1996).

59   "To be gravely affected, one does not necessarily have to drink for a long time": *Alcoholics Anonymous* (Alcoholics Anonymous World Services, Inc., 2001).

61   "More women are drinking, and the women who drink are drinking more": "Gender Bender," *New York,* December 7, 2008.

61   Then, in July 2009, Diane Schuler made headlines: http://www.thefix.com
      /content/"drinking-mom"-syndrome?page=all; http://www.nydailynews.com
      /new-york/driver-deadly-taconic-crash-diane-schuler-drunk-marijuana
      -system-article-1.394195; http://www.nytimes.com/2009/07/28/nyregion/28crash
      .html?pagewanted=all&_r=0.

61   "A Heroine of Cocktail Moms Sobers Up": *New York Times*, August 14, 2009;
      http://www.mommytracked.com/stefanie_wilder_taylor_becoming_sober.

62   "Alcohol is glamorized in our society": Interview with Stefanie Wilder-Taylor,
      January 2011.

63   three different promotional inserts: LCBO promotional supplements, 2011.

64   Clos LaChance, makers of a wine called MommyJuice: http://www.nytimes
      .com/2011/04/24/weekinreview/24grist.html?adxnnl=1&adxnnlx=1368907459-
      wbPUnQgTT9bRwscN9dManw.

64   Why MommyJuice?: Interview with Cheryl Murphy Durzy, May 2011.

64   In Canada, the makers of Girls' Night Out wines: Interview with Doug Beatty,
      2011.

65   Skinnygirl Cocktail line products: http://www.nbcnews.com/business/skinny-
      girl-cocktails-are-fastest-growing-liquor-brand-report-says-746584.

65   musician Fergie of the Black Eyed Peas: http://www.businessinsider.com
      /hanging-out-poolside-with-fergie-at-her-new-vodka-launch-2012-3?op=1.

66   German liquor company G-Spirits: *Toronto Star*, October 5, 2012.

66   Belvedere Vodka: http://atomictango.com/2008/04/29/belvedere/.

66   Jernigan has spent his career watching the industry: Interview with David
      Jernigan, April 2011.

67   the Smirnoff brand: James F. Mosher, "Joe Camel in a Bottle," *American
      Journal of Public Health*, http://ajph.aphapublications.org/doi/abs/10.2105
      /AJPH.2011.300387.

68   "Smirnoff is the girls' vodka": Interview with Kate Simmie, November 2010.

68   "We cannot discount Carrie Bradshaw": Interview with David Jernigan, April
      2011.

69   special taxes on alcopops: https://www.mja.com.au/journal/2011/195/2
      /alcopops-tax-working-probably-yes-there-bigger-picture; interview with David
      Jernigan, April 2011.

69   the Australian Medical Association censured Facebook: "Social Media Con-
      demned for Alcohol Marketing," *Sydney Morning Herald*, September 20, 2012.

69   More than three-quarters of twelve- to seventeen-year-olds in the United States
      own cell phones: http://news.kron4.com/news/more-than-three-quarter-of-
      american-teens-have-cell-phones/.

70 Tea Partay . . . has had more than six million YouTube viewers: http://www
.youtube.com/watch?v=PTU2He2BIc0.

70 There's a strong public health interest in delaying the onset of drinking: Interview with David Jernigan, February 2013.

70 "alcohol could erase pain": Jean Kilbourne, *Can't Buy My Love: How Advertising Changes the Way We Think and Feel* (Simon & Schuster, 1999).

71 "It took a very long time with tobacco": Interview with Jean Kilbourne, January 2013.

CHAPTER 5: THE AGE OF VULNERABILITY

73 "The emptier we feel": Jean Kilbourne, *Can't Buy My Love: How Advertising
Changes the Way We Think and Feel* (Simon & Schuster, 1999).

74 seventeen-year-old Laura: Interviews with Laura, November 2010 and January
2011.

76 Steubenville, Ohio, rape: CNN, March 13, 2013; *New York Times*, March 17,
2013.

76 Audrie Pott: *Associated Press*, April 11, 2013.

77 Rehtaeh Parsons: http://www.thespec.com/news-story/2551427-rehtaeh-parsons-
a-family-s-tragedy-and-a-town-s-shame/.

77 a young woman I will call Rebecca: Interview with Rebecca, February 2013.

78 "If you drink before age fifteen": Interview with David Jernigan, April 2011.

78 "Kids who start early are just different": Interview with Richard Grucza, September 2012.

79 odds of a teenager getting drunk repeatedly are twice as great if they
see their parent under the influence: http://www.dailymail.co.uk/health
/article-2004540/Children-parents-drink-likely-binge.html; http://www.bbc.co.uk
/news/health-13779834.

79 Ask Ali, twenty-two, why she started drinking: Interview with Ali, September
2012.

79 allowing their third-grader children an alcoholic beverage: http://healthland
.time.com/2012/09/20/should-children-be-allowed-sips-of-mommys-drink/.

79 "Young teenagers do not drink with their families at the table the way we did":
Interview with Tiziana Codenotti, December 2010.

80 "Guys drink for the buzz": Interview with Andrew Galloway, September 2012.

80 "It's easy to capture the trends": Interview with Elizabeth Saewyc, April 2011.

81 "we know that genes play a strong role": Interview with David Goldman, September 2012.

82 "there's a lot of sexual trauma": Interview with Brenda Servais, May 2011.

82   "If they're college age, sexual assault is the norm": Interview with Maggie Tipton, June 2011.

82   "their use of alcohol is not to party": Interview with Janice Styer, June 2011.

83   Runaway Intervention Project: Interview with Elizabeth Saewyc, October 2012.

84   Jenna Marbles: http://www.nytimes.com/2013/04/14/fashion/jenna-marbles .html?pagewanted=all; http://www.youtube.com/watch?v=MXzwAXzUwwE.

84   Jennifer Lawrence: http://www.youtube.com/watch?v=CLKZblwLmAY.

85   Which makes it tough to be Melanie: Interview with Melanie, March 2013.

CHAPTER 6: BINGE

87   Let me introduce you to Maggie: Interview with Maggie, August 2012.

89   "University is the acceleration of drinking": Interview with Kate Simmie, November 2010.

90   "The amount of vodka we drank was overwhelming": Interview with Sydne Martin, November 2012.

90   "a handle of vodka for pregaming": Interview with Dana Meyer, June 2012.

91   "drinking efficiently": Interview with Sharon Wilsnack, November 2010.

91   "Their agenda is to get drunk fast": Interview with Ann Kerr, June 2011.

91   "Most girls drink to fit in": Interview with Martha, November 2011.

91   drunkorexia:http://www.sciencedaily.com/releases/2011/10/111017171506.htm; http://www.news.com.au/national-news/drunkorexia-girls-starving-to-drink-more/story-e6frfkvr-1226180554708; http://dailyfreepress.com/2011/11/14/the-skinny-on-drunkorexia/.

92   "University is the age of risk for all psychiatric illnesses": Interview with Kay Redfield Jamison, April 2011.

92   one in four students who showed up at campus health clinics: http://www .ctvnews.ca/many-depressed-students-getting-missed-study-1.599598.

92   students with diagnosed mental health illnesses are the fastest-growing group with disabilities: Interview with Mike Condra, November 2012.

93   Zara Malone: http://www.huffingtonpost.co.uk/2012/10/02/exeter-university-student-drank-herself-to-death_n_1931783.html.

93   Gavin Britton: http://www.dailymail.co.uk/news/article-419496/Student-drunk-death-pub-crawl.html.

93   Nicole Falkingham: http://www.dailymail.co.uk/news/article-2269478 /Nicole-Falkingham-Millionaire-architects-love-split-wife-freezes-death-friends-car-wine-binge.html.

93   "drinking to black out": Interview with David Rotenberg, June 2011.

93 "Girls will bargain their recovery": Interview with Janice Styer, June 2011.

94 "Vodka is very popular for girls": Interviews with Rachel Shindman, October 2012 and February 2013.

94 "People pre-drink because it's cheaper to drink at home": Interviews with Mike Condra, October 2011 and November 2012.

95 the "grease pole": http://robburke.net/greasepole/LegendWeb/Legends /History/index.htm.

96 "No Means Harder": http://www.queensu.ca/news/alumnireview/no-now-really-does-mean-no.

97 Aberdeen Street party: http://www.genx40.com/article/stuff/favoritereading /kingstonwhig26.

98 Habib Khan: http://www.cbc.ca/news/canada/toronto/story/2010/12/03 /ottawa-student-killed.html.

98 "address its culture of drinking": http://www.thestar.com/news /canada/2011/05/31/coroner_rules_alcohol_played_factor_in_deaths_of_ two_firstyear_queens_students.html.

99 an estimated 1,825 Americans die: http://www.ncbi.nlm.nih.gov/pmc/articles /PMC2701090/.

99 one in three college students meets the criteria: Toben F. Nelson and Ken C. Winters with Vincent L. Hyman, *Preventing Binge Drinking on College Campuses: A Guide to Best Practices* (Hazelden 2012).

99 Hospitalizations for alcohol overdoses: http://www.ncbi.nlm.nih.gov /pubmed/21906505.

99 "more young women are binge drinking": Interview with Andrew Galloway, September 2012.

99 "I am not sure they are going to 'mature' out of it": Interview with Sharon Wilsnack, November 2010.

100 "With flavored vodka, the drinking of female college students became much deadlier": Interview with Thomas Workman, May 2013.

100 Samantha Spady: http://compelledtoact.com/Tragic_listing/Spady.htm.

101 how *do* you change a campus culture?: Interview with Rob Turrisi, October 2012.

103 "a targeted motivational intervention for students": Interview with Brian Borsari, October 2012.

103 NCHIP: Interview with Lisa Johnson, May 2013; http://www.nchip.org/.

104 "I'd screen everybody for high-risk drinking": Interview with Michael Fleming, 2012.

CHAPTER 7: SEARCHING FOR THE OFF BUTTON

107 "The central question isn't 'What's wrong with this woman?'": Interview with Nancy Poole, October 2010.

107 childhood sexual abuse or adult abuse histories: Nancy Poole and Lorraine Greaves, eds., *Becoming Trauma Informed* (Centre for Addiction and Mental Health, 2012).

107 majority of young people . . . in the Youth Addiction and Concurrent Disorder Service: Nancy Poole and Lorraine Greaves, eds., *Becoming Trauma Informed* (Centre for Addiction and Mental Health, 2012).

108 "Women are more likely . . . to drink to get rid of negative feelings": Interview with Pamela Stewart, May 2013.

108 what we now call PTSD: http://www.pbs.org/wgbh/pages/frontline/shows/heart/themes/shellshock.html.

110 "The unseen problems of high-functioning professional women are serious": Interviews with Lisa Najavits, February 2011 and October 2012.

118 "If someone steps in and protects you": Interview with Pamela Stewart, May 2013.

118 In Karin's case, no one stepped in: Interview with Karin, July 2012.

120 "You want to know about my drinking?": Interviews with Beata Klimek, May 2011 and June 2012.

CHAPTER 8: SELF-MEDICATION

125 "The breeze at dawn": Coleman Barks, *The Essential Rumi* (HarperCollins, 1995).

128 Women are 70 percent more likely to experience depression: http://www.nimh.nih.gov/statistics/pdf/ncs-r_data-major_depressive_disorder.pdf.

128 a larger number of women have bipolar II: http://healthland.time.com/2011/04/13/catherine-zeta-jones-seeks-treatment-for-bipolar-ii-disorder/.

128 In the United States, a woman is almost 50 percent more likely to walk out of a doctor's office with a prescription: Interview with Susan E. Foster, May 2013.

129 "Couple pills with alcohol, and it's a slippery slope": Interview with Cheryl Knepper, May 2013.

129 Alex was one who used alcohol to treat full-blown bipolar I: Interview with Alex, June 2012.

134 Marion Kane's story: Interview with Marion Kane, January 2013.

137 "With me, it always comes back to anxiety": Interview with Julie, July 2012.

CHAPTER 9: ROMANCING THE GLASS

141 "There is some kiss": Coleman Barks, *The Book of Love: Poems of Ecstasy and Longing* (HarperOne, 2003).

143 "The spider, dropping down from twig": "Natural History" by E. B. White, reprinted in Scott Elledge's *E. B. White: A Biography* (W. W. Norton and Company, 1984).

150 the founder of Women Who Whiskey: Interviews with Julia Ritz Toffoli, June 2012 and May 2013.

152 Scout has been sober—this time—for two and a half years: Interview with Scout, November 2012.

155 Alexandra is an actress: Interview with Alexandra, August 2012.

CHAPTER 10: THE MODERN WOMAN'S STEROID

159 "I can bring home": http://www.youtube.com/watch?v=jA4DR4vEgrs.

162 a recent poll done by Netmums in Britain: http://www.netmums.com/woman/fitness-diet/mums-and-drinking/netmums-and-alcohol-the-full-results/netmums-alcohol-full-survey-results.

162 Jungian analyst Jan Bauer: Interview with Jan Bauer, February 2013.

162 "Perfectionism": Interview with Leslie Buckley, March 2013.

163 a benzodiazepine called Valium: http://www.benzo.org.uk/valium3.htm; http://www.clarkprosecutor.org/html/substnce/prescription.htm.

163 In 1978, it was estimated that a fifth of American women were taking "mother's little helper": National Center on Addiction and Substance Abuse at Columbia University, *Women under the Influence* (Johns Hopkins University Press, 2006).

166 "I suppose I could have stayed home and baked cookies": Hillary Clinton quote at http://abcnews.go.com/blogs/politics/2012/04/working-moms-first-ladies-and-recalling-hillary-clintons-cookies/.

166 I performed the Christmas triathlon: Ann Dowsett Johnston, "The Christmas Triathlon," *Maclean's*, December 10, 2001.

167 "It is my opinion that Norman Rockwell and his ilk": Laurie Colwin, *More Home Cooking: A Writer Returns to the Kitchen* (HarperCollins Publishers, 1993).

167 "I drink until I'm comfortably numb": Interview with Danielle Perron, April 2013.

168 "Partying was really a rite of passage": Interview with Paige Cowan, January 2012.

170 Signing up to be conscious: Interview with Lisa, December 2012.

172 "Alcohol made me social": Interview with Jennifer, September 2012.

CHAPTER 11: THE LAST TABOO

173 "This may be the most stigmatized area": Interview with Janet Christie, March 2011.

173 In 2010, a widely reported British study: http://www.guardian.co.uk /lifeandstyle/2010/oct/06/pregnancy-light-drinking-no-harm-study.

174 "The U.K. study was unfortunate": Interview with Sterling Clarren, November 2011.

175 "the territory is gray": Interview with Nancy Poole, November 2011.

176 "drinking off the charts": Interview with Gerald Thomas, April 2013.

177 "There is no amount of alcohol that has been proven to be safe during pregnancy": Interview with Svetlana Popova, April 2013.

177 "He was the child who would not turn off": Interview with Lynn Cunningham, May 2013.

178 "If you're white, you're ADHD": Interview with Jan Lutke, May 2013.

178 the numbers out of the United Kingdom are high: http://www.telegraph.co.uk /news/uknews/2015160/Half-of-women-drink-alcohol-while-pregnant.html.

179 one in thirteen American pregnant women said that they drank: http://www .cdc.gov/ncbddd/fasd/data.html.

179 a 2012 study in Australia: http://www.fare.org.au/2012/02/2461/.

179 "How do you get over the middle- and upper-class arrogance around FASD?": Interview with Phillip May, December 2012.

181 "Alcohol consumed by the mother passes easily into her breast milk": Interview with Svetlana Popova, April 2013.

182 "We are looking for a marker": Interview with Sarah Mattson, May 2013.

182 "I binge drank through my pregnancy": Interviews with Janet Christie, March 2011 and November 2012.

186 Margaret Leslie and her team . . . at Breaking the Cycle: Interview with Margaret Leslie, July 2011.

CHAPTER 12: THE DAUGHTERS' STORIES

187 "Man hands on misery to man": Philip Larkin, *Collected Poems*, Anthony Thwaite, ed. (Farrar, Straus and Giroux, 2004).

187 "Serious drinkers are like serious eaters": Marion Woodman, *Addiction to Perfection: The Still Unravished Bride* (Inner City Books, 1982).

192 We're all good daughters: Robert Ackerman, *Perfect Daughters: Adult Daughters of Alcoholics* (Health Communications, 1989).

193 daughters raised by alcoholic mothers have different issues than those with alcoholic fathers: Interview with Robert Ackerman, March 2013.

193 "I didn't realize it wasn't normal for a mother to sleep a lot in the day": Interview with Peggy McGillicuddy, December 2012.

195 "My mother didn't want to be here": Interview with Caroline, November 2012.

197 "I refuse to be like my mom—crippled, a depressive": Interview with Kate, December 2012.

198 "She was not a functioning alcoholic": Interviews with Vera Tarman, December 2012 and January 2013.

CHAPTER 13: IN WHICH EVERYTHING CHANGES

205 "If you can kill the right thing": Robert A. Johnson, *She: Understanding Feminine Psychology* (HarperPerennial, 1989).

217 *"I said to my soul, be still"*: T. S. Eliot, *Four Quartets* (Harcourt, 1943).

217 "One of the greatest difficulties in dealing with food addicts, as with alcoholics": Marion Woodman, *Addiction to Perfection: The Still Unravished Bride* (Inner City Books, 1982).

CHAPTER 14: BREAKING THE TRAUMA CYCLE

221 "The past is never dead": William Faulkner, *Requiem for a Nun* (Random House, 1951).

221 "Alcohol was my first experience of getting high": Interviews with Annie Akavak, September 2011 and February 2013.

222 "Braiding" is the approach used at Jean Tweed: Interviews with Nancy Bradley, October and November 2010; November 2011.

223 "Trauma shuts down the frontal lobe": Interview with Susan Raphael, September 2012.

223 "We miss the biggest part of the story": Interviews with Nancy Poole, October 2010 and January 2013.

224 stresses the correlation between trauma and addiction: Interview with Johanna O'Flaherty, December 2013.

224 Jean Tweed began offering child care: Interview with Nancy Bradley, November 2011.

225 "I stopped relapsing": Interview with Annie Akavak, November 2012.

226 "The Children's Program is about breaking the cycle": Interviews with Jerry Moe, December 2012 and February 2013.

228 "There's not much denial with little kids": Interview with Peggy McGillicuddy, December 2012.

229 helping young children make sense of what they remember of your drinking past: Interview with Lesley, March 2011.

230 "It's really difficult for us to forgive ourselves": Interview with Heather Amisson, July 2011.

233 "Let's say there's a frog pond": Interview with Dan Reist, October 2010.

233 "This is a huge public health problem": Interview with Robert Brewer, February 2013.

234 "Lots of harms are coming from those who are not addicted": Interview with Rob Strang, June 2011.

234 "Alcohol is where tobacco was forty years ago": Interview with Mike Daube, October 2012.

235 "When you consider the science, alcohol is doing the most harm in our society": Interview with Jürgen Rehm, May 2011.

236 "Alcohol consumption creates more harm to others than secondhand smoke": Interview with Jürgen Rehm, February 2013.

236 Australian researchers estimated that the costs of harm to others matched the traditional costs of the drinker to society: http://lastdrinks.org.au/research/harm-to-others-summary.pdf.

236 "what we're looking to do is to prevent the early uptake and heavy drinking of young women": Interview with Sally Casswell, April 2013.

237 "Women arguably have the most to gain from a strong policy environment": Interview with Tim Naimi, April 2013.

237 "We don't have the power base to counter the strategic lobbying of the alcohol industry": Interview with James Mosher, April 2013.

238 a 10 percent rise in the price of alcohol is associated with a 9 percent drop in hospital admissions for acute alcohol-related issues: Interview with Tim Stockwell, April 2013.

238 Scotland has indeed declared its intention to set a minimum price for a standard unit of alcohol: http://www.bbc.co.uk/news/uk-scotland-scotland-politics-22394438.

238 Minimum pricing: http://www.ias.org.uk/newsroom/uknews/2013/news010513/iasreport-stockwell-thomas.pdf.

238 Prime Minister David Cameron seems to have reneged on his promise: http://www.telegraph.co.uk/health/healthnews/9987635/Pubs-demand-minimum-alcohol-price.html.

239 more than a million alcohol-related hospital admissions each year: http://www.ias.org.uk/newsroom/uknews/2013/news300513.html.

239 Federal alcohol taxes have not been raised since 1991: Email exchange with David Jernigan, April 2013.

239 "Ultimately, cheap booze is not healthy for society": Interview with Tim Naimi, April 2013.

239 "gender convergence is propping up sales": Interview with Thomas Greenfield, April 2013.

240 alcohol is 44 percent more affordable, in real terms, than it was in 1980: http://www.dailymail.co.uk/news/article-2235017/Low-price-alcohol-mean-youngsters-drink-activities.html.

240 alcohol companies have discovered they need women to drink—and they are marketing accordingly: Interview with David Jernigan, February 2013; and email exchange, April 2013.

241 "FAS or FASD is just one example of harm to others": Interview with Jürgen Rehm, February 2013.

242 "Addiction is not a moral failure": Interview with Susan E. Foster, May 2013.

242 random breath testing is the future: Interviews with Andrew Murie, 2010, 2011, 2013.

243 "we need to understand is the coexistence of depression and drinking in women": Interview with Nancy Poole, February 2013.

244 "Becky's Not Drinking Tonight": http://www.youtube.com/watch?v=hIe9KGAJeko.

244 "The highest risk is for the higher-educated women in lower-resourced countries": Interview with Sharon Wilsnack, March 2011.

CHAPTER 16: WRESTLING WITH THE GOD THING

245 "The spiritual life is not a theory": *Alcoholics Anonymous* (Alcoholics Anonymous World Services Inc., 2001).

250 *Spiritus con spiritum*: Marion Woodman, *Addiction to Perfection: The Still Unravished Bride* (Inner City Books, 1982).

250 Jung corresponding with Bill Wilson: Jan Bauer, *Alcoholism and Women: The Background and the Psychology* (Inner City Books, 1982).

251 Jung "recognized the confusion between physical and spiritual thirst": Marion Woodman, *Addiction to Perfection: The Still Unravished Bride* (Inner City Books, 1982).

251 the phrase "God as we understand Him": http://www.barefootsworld.net/aadilemmanoffaith.html.

251 "We live in a predominately Christian culture": Marion Woodman, *Addiction to Perfection: The Still Unravished Bride* (Inner City Books, 1982).

253 "Giving up booze": Wilfrid Sheed, *In Love with Daylight: A Memoir of Recovery* (Simon & Schuster, 1995).

254 "flashy tail feathers of the bird courage": E. B. White, as quoted in Scott Elledge's *E. B. White: A Biography* (W. W. Norton and Company, 1984).

254 "a bunch of sober alcoholics": Anne Lamott, *Travelling Mercies: Some Thoughts on Faith* (Anchor, 1999).

255 One of the sober alcoholics who helped me stay sober is Claire: Interview with Claire, January 2013.

258 "Addiction attempts to fill a spiritual void—and of course, it doesn't work": Interview with Stephanie Covington, December 2012.

260 "I was a distant stranger to my kids": Stephanie Covington, *A Woman's Way Through the Twelve Steps* (Hazelden, 1994).

261 "The word God could have a religious meaning": Gabor Maté, *In the Realm of Hungry Ghosts* (Alfred A. Knopf Canada, 2008).

261 "Underneath the personality, there's a deeper self": Interview with Gabor Maté, February 2013.

262 "This is a disease of perception": Interview with Karin, January 2013.

CHAPTER 17: STIGMA

263 "It's the misunderstanding that kills": Interview with Susan Cheever, June 2011.

263 "You could say to me, 'Drink responsibly,' and I'll say: 'I'll try!'": Craig Ferguson at http://www.youtube.com/watch?v=kI-BhQGwDO8.

264 "When the series is over, will you need a job?": Interview with Alison Uncles, November 2010.

266 "We have known for more than twenty years that alcoholism is a chronic, relapsing brain disease": Interview with Peter Thanos, November 2011.

266 "The American Medical Association calls it a disease": Interview with Patrick Smith, November 2011.

266 "Our only shot is to see it as a disease because there's much less stigma": Interview with Jean Kilbourne, March 2010.

266 "If, in fact, alcoholism is a kind of slow suicide": Jan Bauer, *Alcoholism and Women: The Background and the Psychology* (Inner City Books, 1982).

268 "Trauma isn't just an emotional event: it shapes brain circuitry": Interview with Gabor Maté, February 2013.

268 "For a woman, the stigma of addiction is still very bad": Interview with Sheila Murphy, May 2011.

268 "no one *chooses* to be an addict": Interview with Stephanie Covington, April 2013.

268 "A big process of my recovery was self-forgiveness": Interview with Rebecca, March 2013.

268 "I wish I could use my real name": Interview with Scout, November 2012.

269 Is it time for AA to drop the second A?: *New York Times*, May 6, 2011.

270 "We are in the midst of a public health crisis": Susan Cheever at http://www .thefix.com/content/breaking-rule-anonymity-aa.

270 "Post-treatment, those who remain involved in a mutual twelve-step program like AA dramatically improve their chances of remaining clean and sober": Interview with Patrick Smith, June 2011.

270 "if we cross into addiction, there is no going back": Interview with Johanna O'Flaherty, November 2012.

271 "Anonymity protects": Interview with Susan Cheever, June 2011.

271 "I think that the recovery world is where the gay world was in the 1990s": Interview with Maer Roshan, June 2011.

271 "I do think recovery needs a face and a voice": Interview with Stephanie Covington, July 2011.

272 "the hostility to meetings always surprises me": Interview with Susan Juby, June 2011.

272 "If you're female and drunk, you're not going to live it down": Interview with Karin, June 2011.

272 "One of the greatest problems is that the general public and policy makers don't understand that people *can* and *do* get well": Interview with Pat Taylor, June 2011.

272 "Recovery is a mind, body, and soul experience": Interview with William C. Moyers, June 2011.

273 "you don't need alcohol to be confident": Interview with Chris Raine, April 2013.

274 National Roundtable on Girls, Women, and Alcohol: http://www.thestar.com /news/gta/2013/03/08/sobering_lack_of_progress_seen_on_women_and_ alcohol.html.

CHAPTER 18: BECOMING WHOLE

275 "You will lose someone you can't live without, and your heart will be badly broken": Anne Lamott, *Plan B: Further Thoughts on Faith* (Riverhead Books, 1995).

275   For me, there has been one central fairy tale that has been totemic in my jour-
      ney: Marie-Louise von Franz, *The Feminine in Fairy Tales* (Shambhala, 1993);
      and Robert A. Johnson, *The Fisher King and the Handless Maiden: Understanding
      the Wounded Feeling Function in Masculine and Feminine Psychology* (Harper-
      SanFrancisco, 1993).

280   "For the Breakup of a Relationship": John O'Donohue, *To Bless the Space
      Between Us: A Book of Blessings* (Doubleday, 2008).

282   "You are standing on the shaky sands of doubt": Fraser Boa at http://blog
      .gaiam.com/quotes/authors/fraser-boa.

ABOUT THE AUTHOR

The winner of five National Magazine Awards, Ann Dowsett Johnston is a writer and editor recognized for her expertise in higher education and alcohol policy. A recipient of an Atkinson Fellowship in Public Policy and a Southam Fellowship in Journalism, she spent most of her professional career at *Maclean's* magazine, where she was best known as the chief architect of the university-rankings issue. A graduate of Queen's University and a former vice principal of McGill University, she lives in Toronto.